Origins and Futures:
Time Inflected and Reflected

The Study of Time

Founding Editor

J.T. Fraser

VOLUME 14

The titles published in this series are listed at brill.com/stim

Origins and Futures: Time Inflected and Reflected

Edited by

Raji C. Steineck and Claudia Clausius

BRILL

LEIDEN • BOSTON
2013

Cover illustration: Vienna, Schönbrunn gardens, statue Janus and Bellona. http://commons.wiki media.org/wiki/File:N29Janus-u-Bellona.jpg, author: schurl50.

This publication has been typeset in the multilingual "Brill" typeface. With over 5,100 characters covering Latin, IPA, Greek, and Cyrillic, this typeface is especially suitable for use in the humanities. For more information, please see www.brill.com/brill-typeface.

ISSN 0170-9704
ISBN 978-90-04-25168-7 (hardback)
ISBN 978-90-04-25200-4 (e-book)

This book is printed on acid-free paper.

CONTENTS

Obituary ix
List of Contributors xi
Acknowledgements xv

Introduction 1
Raji Steineck and Claudia Clausius

SECTION I

ORIGINS EXPLAINED

Founder's Lecture: The Origin of the Integrated Study of Time 15
J.T. Fraser

The Search for Origins as Detective Story: Detecting Datelines and the Mystery of Origins 25
Paul A. Harris

On the Social Origin of Time in Language 37
Walter Schweidler

The Origins of Language and Narrative Temporalities 49
Rosemary Huisman

SECTION II

FROM FUTURE TO ORIGIN:
TEMPORAL INVERSIONS AND ASYMMETRIES

Narrative Fiction: Writing towards the Origin 79
Sabine Gross

Origins as Futures in the Time Plays of J.B. Priestley 103
Carol Fischer

Big Science: Marching Forward to the Past 123
Michael Crawford

The Past-Future Asymmetry 139
Friedel Weinert

SECTION III

Futures Opened, Futures Closed

The Human Temporal Condition between Memory and Hope 169
Steven Ostovich

Bachelard's "Discontinuous Bergsonism" in Zora Neale Hurston's "Sweat": How Self-Generation of "Pure Time" Engenders Free Choice 189
Patricia McCloskey Engle

Tales of Time and Terror: Walter Benjamin's Philosophy of History and the Narrative Aesthetics of Edgar Allan Poe 207
Marcus Bullock

The Future of August 6th 1945: A Case of the 'Peaceful Utilization' of Nuclear Energy in Japan 235
Masae Yuasa

Forgiveness as the Opening of the Future 259
Karmen MacKendrick

Index 281

Julius T. Fraser (1923–2010)

OBITUARY

Julius T. Fraser died November 20th, 2010 in Westport, Connecticut, and a great number and diversity of people internationally mourned his passing. Professor Frederick Turner (University of Texas, Dallas) summed him up well: "J.T. Fraser is arguably the most important philosopher of the twentieth century. Like Aquinas, and like Aristotle before him, he has created the very arena within which philosophy will be conducted for the next few decades, perhaps centuries." In a review of one of his books, the *Times Higher Education Supplement* in 1990 declared: "The conceptual apparatus Fraser has built up is . . . of almost unprecedented breadth." Professor George M. O'Har (Boston College) commented regarding Fraser's book *Time, Conflict, and Human Values*, "the book is a treasure trove of imaginative scholarship and wisdom of a kind so rarely seen today as to be almost extinct."

Fraser was born in Budapest, Hungary on the 7th of May, 1923. His ancestry made for difficult times during the Second World War, but with the collapse of the Nazis life did not become any easier. He managed to escape to the west from a Soviet labour brigade, and he eventually arrived penniless in New York City. Working at whatever he could lay hands to, he eventually trained as an engineer in the United States where his successful patents on microwave devices allowed him to complete a PhD at the age of 40 at the University of Hannover, Germany. Fraser was remarkable for more than just his intellect and drive: striving to make sense of his experiences during the war, he came to realize that all human life, philosophy, and endeavor could be understood through an analysis of temporal experience. This led him to assemble a framework within which diverse intellectual traditions could converse productively. He could only have constructed his *Hierarchical Theory of Time* by becoming a master of hugely disparate fields—to do so he came to exemplify the sort of polymath Natural Philosopher that has not been seen since the turn of the last century when physics, biology, cosmology, etc., were considerably simpler and more tractable. In the mid 1960s he founded the International Society for the Study of Time, which gives the term "interdisciplinarity" a head-stretching workout. The Society has acted as a nexus for vigorous and challenging debate among broad-minded philosophers, physicists, social scientists, literary critics, psychologists, historians, biologists, judges, authors, artists, musicians, etc. . . . ever since.

Despite his intimidating intellect "JT," as he was known to his friends, was a small man, and blue eyes penetrated playfully from beneath a brown bald head. What struck you most upon meeting him was his warmth, generosity, and humanity. How many academic careers has this man helped to nurture and launch? His good humour and mellifluous voice could lull you into a false sense of security before the spark of his riveting and broad-ranging conversation struck. If you were lucky, you found yourself challenged, taxed, and happily exhausted from the workout. He enjoyed his colleagues and friends, and traveled extensively with his indefatigable and engaging wife Jane, lecturing and conferencing his way across the globe. While he was ever the cultural sophisticate, he was not above putting the after-dinner tea cozy on his head to amuse the children while discussing Kierkegaard, Malthus, or Hawking with his hosts. If there were a Nobel Prize for performing a vitally integrative function, the cross-disciplinary discourse this man ignited would have won him recognition in an instant.

He had declined rapidly over the last year of his life, and his body was no longer able to keep measure with his coursing mind. A huge intellect and a deep soul, the greatest Timesmith of them all passed quietly in the company of his wife and family. We are the poorer for the loss.

they are, it was clear to all—and he was the first to point it out—that sooner or later, ISST would have to continue without him. In fact, the 2010 conference became the last one to have a "Founder's Lecture," one which had to be delivered by his long-time friend Frederick Turner, because J.T. Fraser was already too frail to make the journey, passing away only months later at his home in Westport, CT.

The selection of the theme "Origins and Futures" and of the Monteverde conference venue reflected a desire to connect to origins and think about possible futures—not only for ISST, but also for humanity in general, which many of us perceive to stand at the crossroads leading either to self-induced extinction or to the development of a sustainable economy and society. However, in a move typical for the inquisitive spirit of the Society, the conference went beyond the pattern of looking to a fixed past in order to extrapolate a pre-determined future. The title's plural forms were a deliberate call for open reflection, not only on various points of departure and origination and models of and for futurity, but also on the complex relationship between origins and futures. The Founder's lecture (presented here in an edited version) traced the origin of J.T. Fraser's sustained search for an integrated theory of time, a theory attempting to accommodate the whole range of time felt and time understood. In relating the story of a young man caught between two armies trying to force their final solution on the world as the origin of his quest, he shared a decisive part of his life's experience—and, at the same time, made the audience aware how vastly different origins can still lead to common interests and goals.

This sharing of zeal and interest crystallized in several papers critically engaging with Fraser's theory of time. Literary scholar Rosemary Huisman uses his theory of a nested hierarchy of levels of causality and temporal ordering to explore various modes of narrative prose. Steven Ostovich discusses his notion of sociotemporality as a distinct level of temporal ordering and coherence above "nootemporality" (the temporal world of the individual *homo sapiens*). Walter Schweidler's philosophical reflection on the "Social Origin of Time in Language," although not explicitly referring to Fraser, can be read in a similar vein, as postulating an integration of socio- and nootemporality, or, to put it differently, their co-origination, instead of sociotemporality emerging out of nootemporality. The creative manner in which Fraser's concepts are employed, and their problems discussed, demonstrates that his theory, important as it continues to be as a point of reference for many members of ISST, is far from being treated as doctrine or dogma. This stimulus also exemplifies the spirit of debate

LIST OF CONTRIBUTORS

MARCUS BULLOCK is Professor Emeritus in the English Department of the University of Wisconsin, Milwaukee, and Research Associate in Comparative Literature at UW-Madison. He collaborated as co-editor on the Harvard edition of Walter Benjamin's *Selected Writings* and has published books or essays on Walter Benjamin, Ernst Jünger, Franz Kafka, Friedrich Schlegel, Alfred Andersch, Dante Gabriel Rossetti, Julia Kristeva, Jacques Derrida, and Martin Heidegger.

CLAUDIA CLAUSIUS (PhD) received graduate degrees from Oxford University and the University of Toronto. Currently, she is Associate Professor and Chair in the Department of Modern Languages at King's University College at Western University, Canada. Her on-going research interests focus on the intersection between modern art and contemporary drama. She is currently also editor of *KronoScope: Journal for the Study of Time.*

MICHAEL CRAWFORD (PhD) is Professor of Biology at the University of Windsor in Canada. His research interests focus upon embryology, gene regulation, and more recently, upon the role of temporal pressures at the interface of scientific enterprise and public perception. Like the Human Genome Project, he is past middle age and finds himself looking retrospectively with an element of delight and surprise . . .

PATRICIA MCCLOSKEY ENGLE, PhD, teaches in the Language and Literature Department at Bucks County Community College. Her scholarly work has appeared in *DQR Studies in Literature, Explicator*, and Lehigh University's online *Literature of Justification* project. She is a past editor of the Artsbridge *Writers' Gallery*, and her poetry has appeared in *The River*.

CAROL A. FISCHER has worked in technical theatre arenas for 20 years, assuming the various roles of costume-, light-, sound designer, stage and production manager, technical supervisor, and director. She earned an MA degree in Theatre Arts from San Jose State University (2004), and a PhD in Theatre Studies at UC Santa Barbara (2010). Currently she teaches, directs, and designs at West Valley College in San Jose, CA, USA.

J.T. Fraser (†) was the author of six volumes, editor of eleven, and author of more than seventy articles in professional periodicals on themes related to the integrated study of time. He was Founder of the International Society for the Study of Time and Founding Editor of *KronoScope: Journal for the Study of Time.* He lectured extensively at home and abroad about themes related to the study of time.

Sabine Gross is Professor of German at the University of Wisconsin-Madison, where she teaches literature and theater. She has published widely on literature of the eighteenth- through the twenty-first centuries with an emphasis on contemporary German-language authors; narratology and stylistics; the reading process; image-text relations; the human experience of time; semiotics; film and theater; perception and synaesthesia.

Paul A. Harris is Professor of English at Loyola Marymount University, Los Angeles, where he teaches interdisciplinary courses on themes such as Wonder, Chaos, Time, and Nothing. He served as President of the International Society for the Study of Time from 2004 to 2012 and is co-editor of *SubStance: A Review of Theory and Literary Criticism.* He has published interdisciplinary work on time in philosophers (Bergson, Deleuze, Serres) and writers (Poe, Melville, Faulkner, Perec), and he enjoys writing constraint-based poetry that he performs with jazz musician David Ornette Cherry.

Rosemary Huisman is Honorary Associate Professor in English at the University of Sydney, Australia. Her research publications include *The Written Poem, Semiotic Conventions from Old to Modern English* (London & New York: Cassell 1998 & Continuum 2000) and, as co-author, *Narrative and Media* (Cambridge U.P., 2005). She is past President of ASFLA, the Australian Systemic Functional Linguistics Association.

Karmen MacKendrick is Professor of philosophy at Le Moyne College in Syracuse, New York. Her recent books include *Fragmentation and Memory: Meditations on Christian Doctrine* (Fordham, 2008); with Virginia Burrus and Mark Jordan, *Seducing Augustine: Bodies, Desires, Confessions* (Fordham, 2010), and *Divine Enticement: Theological Seductions* (Fordham, 2013).

Steven T. Ostovich is Professor and Chair of the Philosophy Department at the College of St. Scholastica in Duluth, Minnesota, USA. His interests include political theology, philosophy of science, and time and history, and his most recent publication is the co-authored book *The Happy Burden of History* (De Gruyter 2011). He currently serves on the ISST Council and has been Interim Editor of the ISST's journal, *KronoScope*.

Walter Schweidler holds the Chair of Philosophy at the Catholic University Eichstätt-Ingolstadt since 2009. From 2000 to 2009 he was Professor for Practical Philosophy at Ruhr-University Bochum. His research interests include contemporary and modern concepts of ethics and political philosophy; philosophy of law and human rights theory; phenomenology, Heidegger's philosophy in the context of the main philosophical concepts of the 20th century; metaphysics and the critique of metaphysics; intercultural philosophy; bioethics. Selected publications include *Die Überwindung der Metaphysik*, Stuttgart 1987; *Geistesmacht und Menschenrecht: Der Universalanspruch der Menschenrechte und das Problem der Ersten Philosophie*, Freiburg/München 1994; *Das Unantastbare: Beiträge zur Philosophie der Menschenrechte*, Münster 2001; *Der gute Staat: Politische Ethik von Platon bis zur Gegenwart*, Stuttgart 2004; *Das Uneinholbare: Beiträge zu einer indirekten Metaphysik*, Freiburg 2008; *Über Menschenwürde: Der Ursprung der Person und die Kultur des Lebens*, Wiesbaden 2012.

Raji C. Steineck is Professor of Japanology at the University of Zurich. His research interests combine medieval and contemporary history of ideas in Japan and the philosophy of culture, with a special interest in problems of symbolic expression and time. He currently serves on the ISST Council and has contributed various articles to ISST's scholarly publications, *KronoScope* (e.g. Vol. 3-2: "On 'Brain Death,' Death, and Personal Identity") and *The Study of Time* (Vol. 12: 'The Body as a Medium of Memory"; Vol. 13: "Truth, Time, and the Extended Umwelt Principle: Conceptual Limits and Methodological Constraints").

Friedel Weinert is Professor of philosophy at the University of Bradford in the United Kingdom. He specializes in the study of the interrelations between science, its history and philosophy. He has published and edited several books on these interactions, covering both the natural and the social sciences. He contributed the paper "Temporal Asymmetry and Relativity" to Volume XIII of the Study of Time (2010).

Masae Yuasa is Professor of Sociology at Hiroshima City University. She has written articles on a wide range of subjects, including the political economy of Japanese textile industry, social theories on globalization and identity, and sociology of labor and time. Since 2008 she has explored issues of Japanese nuclear policies and industry. Her current research concern is time and social order/disorder in Japanese society especially after the Fukushima nuclear accident.

ACKNOWLEDGEMENTS

This book was made possible by the efforts of many more people than those listed in it as authors and editors. We would like to thank everyone at the Monteverde conference site for providing such an inspiring and agreeable environment for our meeting.

Tom Weissert created and maintained the volume website. Daniela Tan greatly helped us with copy-editing, indexing and unifying citation formats. Joed Elich guided us smoothly through the publishing process at Brill. We are deeply grateful for their support.

Our profound thanks go to the reviewers who evaluated the manuscripts and gave valuable suggestions for improving them. They include Hervé Barreau, Craig Callender, David Chiavacci, John Downey, Graeme Gilloch, Todd Herzog, Petra Hroch, Remy Lestienne, Jürgen Manemann, Zara Mirmalek, Jo Alyson Parker, Ninette Poetzsch, Anca Pusca, Conrad Russell, and Jonathan Tallant, and others, who have chosen to remain anonymous.

INTRODUCTION

Raji Steineck and Claudia Clausius

When the future seems uncertain, we look back towards origins. In times of crisis, grasping an origin is like identifying a fixed point, providing us with a sense of destiny and direction. Remembering how we came to be what we are also serves to confirm for us the creative forces that shaped us, reinforcing our power to cope with change and create ourselves anew. Individuals and communities of all sizes and shapes, therefore, regularly develop rituals of remembrance: not necessarily out of a desire for regression or nostalgia, but more often as a way to reassure themselves of their power to face the future with the wisdom that helped them to create and maintain a sense of purpose, identity, and direction. We are as fixated on knowing the past as we are on foreseeing the future; all our endeavours equal methods of negotiating within the temporal polarities of beginning and ending. And we are ever vigilant for expressions that promise to reward or justify these wishes. Near the end of her iconic Canadian novel *The Diviners*, Margaret Laurence writes:

> The sun, now low, was catching the waves, sending out once more the flotilla of little lights skimming along the greenbronze surface. The waters flowed from north to south, and the current was visible, but now a south wind was blowing, ruffling the water in the opposite direction so that the river, as so often here, seemed to be flowing both ways.

We all see the current, and we take delight in the sparkle of the waves' surface, but what deeply thrills our senses is the play between wind and water that performs the river's dual directionality.

This volume presents selected papers from the International Society for the Study of Time's XIV triennial international conference, on the theme of "Origins and Futures," held in Monteverde, Costa Rica, amidst a natural rainforest reserve. When the ISST's Council started deliberations on the theme and place for this conference, the Society had clearly entered a phase of transition. Our founder, J.T. Fraser, whose indefatigable energy had propelled ISST on for four decades since its inception in 1966, had withdrawn from much of the daily business of the Society. Although he still participated in the Council's work, infusing it with his wisdom, sense of purpose, and not least with a humor that accepted the facts of life as and dialogue as it impregnated the Monteverde conference and as it, one hopes, is evident in this volume, a spirit marked by openness, interdisciplinarity and a serious engagement with matters of central concern to current human society.

The present volume does not cover the conference as a whole; instead, it presents a selection of papers from a variety of disciplines that coalesce around three common themes: conceptualizations and explanations of "origins" (section 1); the complex, sometimes paradoxical relationships between origins and futures (section 2); and the possibilities and strategies to re-open a future that appears pre-determined by "original experiences' of violence and injustice (section 3). Not surprisingly, additional connections arise between papers across sections, such as conversations on aspects of narrative structuring or dialogues on the dynamics between individual and social time-consciousness (nootemporality and sociotemporality). We hope that one of the pleasures of the volume is the reader's opportunity to further link, pattern, invert, and otherwise reconfigure the papers' various concepts about origins and futures with and against each other.

Section I, "Origins Explained," explores both the notion of origin and specific cases of origination. It starts out with the Founder's lecture, which recounts the personal experiences that led to the foundation of ISST and the formulation of the integrated, hierarchical theory of time. In the ensuing President's address, Paul Harris elucidates two possible ways to conceptualize origins and concomitant ways to explore them: the detection of datelines and the metaphysical reflection on an origin that defies reification. Drawing on one case each from natural and cultural history, he describes how the search for an origin follows the narrative logic of a detective story in the identification and analysis of traces and their possible causal links. This mode of operation, however, runs into a natural boundary where the origin of the universe is concerned. Detection can only take place where data exist that can be interpreted as traces causally linked to the event in question. In the absence of data from the "singularity at point zero," inquiries into the origin of the universe turn, in the words of cosmologist Eric Chaisson, into reflections about the "origin of the origin," and thus switch from "detective" to "mystery" mode. Drawing on classical examples from Augustine to Chuang Tzu, Harris elucidates the paradox such reflection of necessity entails, and turns to the symbolism of the Ouroboros and to a concrete poem by Ronald Johnson to analyze how the concept of an ungraspable origin may nevertheless be lucidly expressed.

Walter Schweidler offers another example of the metaphysical form of reflection on origins in his philosophical search for the origin of the concept of time in human society that does not, however, end up in social constructivism. He builds on Heidegger's observation that the experiential reality of time originates with the existence of *Dasein*, that is, a being who has an understanding of itself, and he turns to Max Scheler's term *Daseinsrelativität* to further elucidate how time is not "constituted by subjective consciousness," but is "ontologically dependent on its existence"—i.e. the existence of "time" can only be grasped within the existence of a conscious being that finds itself as existing "in time." This, Schweidler goes on to say, is only possible in the way that the subject finds itself as *having already* existed and, without explicitly knowing it, having exercised an understanding of its existence, and the primary medium that exemplifies and makes possible this complex structure is language. Language in turn is an essentially social phenomenon, the origin of which, Schweidler maintains, cannot be understood to lie in social convention or construction. Language, therefore, is of necessity perceived by the society that lives through it as given by an originating force external to this society. Thus, while the epistemological origin of time is social, the conceptualization of the origin of this origin leads back to a pre-existing creative force or principle.

In the last paper of this section, Huisman analyzes the origin of a specific but essential use of language in her analysis of narrative temporalities. Drawing on the typology of processes in English developed in Michael Halliday's functional linguistics, she demonstrates how Fraser's evolutionary hierarchy of levels of temporality is to some extent reversed in the development of narrative techniques in English narrative prose. She finds that in Old English epics like *Beowulf*, the narrative, although making use of the chronological sequence typical to biotemporality in the depiction of actions, is dominated by "equative sequences" in which protagonists and actions are presented as examples of certain essential types. Thus, the narration of the thematic plot is interspersed with accounts of analogous events and persons, to the point of frustrating the modern reader's expectations for "narrative advance." In terms of temporality, these epics capture not the timescape of an individual agent, but that of a society. It is only with the classic novel that the representation of individual nootemporality through associative sequences becomes pre-eminent. However, it remains intertwined with chronological and equative sequences relating to the biotemporal and sociotemporal worlds. Huisman continues by claiming that with the extensions of the human umwelt through the progress of the natural sciences, new modes of narrative became conceivable that gave expression to the reversible, indeterminate, or even incoherent sequences of the "eotemporal," "prototemporal," and "atemporal," respectively defined by modern physics' description of inorganic matter, particles, and radiation. Once again, what emerges from Huisman's account of the development of English narrative prose is an "ouroborical" image of origins, where those levels of temporality that we assume to come first in terms of evolution are the last ones to be known and expressed.

Section II, "From Future to Origin: Temporal Inversions and Asymmetries," explores conceptual interdependences and interactions between origins and futures, and the fundamental notion of temporal asymmetry that is the structural prerequisite for the conceptualization of both.

This sections begins with explorations of the various modes of reversal that may occur in the dynamic between an origin and its concomitant future.

Sabine Gross analyzes the complex relations between origins and futures in plot construction and in the reading of fiction, especially detective fiction, in her paper "Endings as Origins." Gross begins by illustrating how all realist texts are examples of a "double logic," where the text's pretence of plausibility hides its own structure, or what one might term an author's deliberate narrative plan, a kind of reverse causality. However, detective fiction in particular defines itself by a self-conscious second-order of writing backwards by requiring readers to read both forwards and backwards and by compelling them to evaluate incoming information in light of both a past crime and its future resolution. Extraordinarily, our suspense and pleasure in the text do not in practice depend exclusively on our ignorance of its ending. Gross argues that suspense is an aesthetic effect in which we commit ourselves to being future-bound—we inhabit a double virtual arc between origin and future. The ways in which writing backwards is made by authors and by literary genres to appear as writing forward run counter to the way readers experience time in their real lives. Detective fiction, Gross claims, exposes the specific pleasure that readers gain by involving themselves in this double logic. The aesthetic experience here seems to position the reader on a plane with that of the author, enjoying similar active powers of deduction, prediction, exposure, resolution, and re-establishing order and stability.

Carol Fischer investigates J.B. Priestley's adaptation of time theories in several of his plays, where precognition of the future profoundly changes the attitude of the protagonists. Where Gross had described drama as one of the key genres involved in writing/reading backwards (prototypically *Oedipus Rex*), Fischer scrutinizes the precise ways in which Priestley's dramas stage characters who miraculously, but believably, become drawn into temporal reconfigurations that allow them to re-decide their lives, thereby altering the determination of those previous existences in both an individual and collective manner. Priestley's dedication to a more caring human community drove him to revolutionize the content of British drama in order to stage, as entertainment, ways in which humans might actively undertake to improve the world through their own personal choices. In a way similar to how Gross endows the reader with strategic authorial textual powers (provided they are alert and experienced readers), Fischer shows how Priestley arranges the drama's characters in various time theories. She points out that both P.D. Ouspensky and J.W. Dunne spatialize time by way of illustrating how humans might penetrate into time's other manifestations. Ouspensky believed that time had up to six dimensions, in the final of which the spiral-like circles of time stretch out so that an individual can—through his or her evolutionary journey—vaguely recall previous lives, and thus alter decisions and actions towards a more beneficial future. J.W. Dunne's serialism also spatializes time; individuals occupy specific observer roles offering them access in certain states (dreaming, for instance) to another level of temporal consciousness. The scientific validity of Priestley's views on time are beside the point here; his keen sense of what makes for engaging theatre and what compels an audience's attention empower him to give life to otherwise esoteric time theories in such a way as to stage them as compelling topics for reflection.

In "Marching Forward to the Past," Michael Crawford proposes a reverse image of Priestley's fascination with direct access to the future. In an appraisal of the history of recent and contemporary "Big Science" endeavours such as the Human Genome Project, he demonstrates how scientific mega-projects that set out to explore the future (or futuristic possibilities) inevitably turn to the past as a source of knowledge and information. In these instances, the initial emphasis is on the potential of the planned research to revolutionize biology and the "future prevention and treatment of illness." In the course of the research, however, it soon becomes evident that the immediacy of benefits for humanity was naively exaggerated. Instead, the project has run into substantial surprises regarding the size, structure, and subtlety of the human genome. Some attention has consequently shifted to focus upon comparative genomics and the evolutionary past. This readjustment of perspective has, nonetheless, provided tools to estimate and even to re-build ancestral genes, and by extension to infer the life habits of *Archosaurs.* Crawford emphasizes the structural necessity that led to this turn from future possibilities to evidence from the past, and adduces the nascent Human Epigenome Consortium to show how the same logic plays out in this, equally grand, follow-up to the Human Genome Project. He infers that the even more futuristic-sounding project of Exobiology will most likely follow a similar trajectory and argues that "the mega-projects of Basic Science are profoundly retrospective no matter what the espoused agenda and temporal orientation." Observing how the challenges posited by such mega-projects have also fostered a spirit of international collaboration, Crawford in conclusion expresses a hope that they may herald a global quest for peaceful knowledge.

From the papers by Gross, Fischer and Crawford, one draws the more general observation that origins are not simply a matter of the past: it is their potency to generate a future that makes them a compelling point of interest—and it is only the future by which we can tell what was an origin. While the origin seems to be the exemplar of something given, its identification involves a choice concerning what the object of explanation is, and how its future may be envisioned.

The final paper in this section, "The Past-Future Asymmetry," explores the physical realities behind a notion central to all conceptualizations of origins and futures: temporal asymmetry. Friedel Weinert carefully assesses specific time theories in physics. His conclusion claims that—concepts of a "block universe" notwithstanding—there is good reason to believe that the asymmetry between past and future events is pertinent already on the level of inanimate matter. The distinction between past and future can therefore be said to be more than a useful illusion, or a mere construct, of the human mind. Weinert playfully but earnestly makes use of literary texts to interrogate the dynamics of time travel into the past and into the future. The article offers a series of charming but mind-bending "stories" from Wells's time traveller, to the twin paradox tale of Homer and Ulysses, to Laplace's demon, to Dummett's story of a painter, to interrogate and test the various versions of time reversal in physics. Taking entropy into account, he concludes that time travel into the future keeps the time reversal asymmetry intact. Time is not simply a spatial trajectory upon which a time visitor travels up and down; it is a sequential distribution of energy states that cannot be undone. This analogy of "spreading" energy, however, offers a future possibility referencing the past as well as a fully potential present. Like the dynamic of forgiveness in Karmen MacKendrick's article, the past can only be altered through a revision of the future. The asymmetry of time that makes travel to the past impossible nevertheless opens up the past by way of an open future, and also introduces the possibility of liberation from determinism via the past.

Section III, "Futures Opened, Futures Closed," is dedicated to the specifically human concern with freedom, which presupposes a future not fully determined by conditions and events of the past. Several papers explore specifically the possibility of "re-opening" the future despite its seeming completely determined by some past trauma or catastrophe.

In his assessment of the human temporal condition, Ostovich brings together Fraser's theory of time as a nested hierarchy of temporalities with Hannah Arendt's notion of natality. His objective is to make sense of the temporal finitude of human existence in a way that avoids both the denial of finitude and a cheap, fatalistic acceptance of things as they are that would preclude all hope for improving human society. Ostovich inherits from Fraser the idea that not only is there no escape for humans from time into eternity, but also that there is, in the final analysis, no good reason to *want* to escape into eternity because, in the words of Fraser, "only the temporal is open to creation and freedom." Creativity is the hallmark of a decidedly temporal existence, one that is aware of origins and characterized by its "capacity to give birth to the new," as Ostovich explains Arendt's term "natality." From this basic assessment, Ostovich moves on to explore, with Arendt, the horizon of human freedom in terms of social action, and the way social action *qua* realization of freedom is linked to public memory, forgiveness, and the hope for justice. With respect to the latter, he subsequently probes into the possibility of an "eschatology without god" that could function to prevent social and political inertia. Eschatological hope, Ostovich maintains, can help us to act "as if the world were not condemned to injustice," and thus make sense of the struggle for justice as a rational response to the tragic quality of human life.

Patricia McCloskey Engle turns to a literary example of breaking away from brutal past conditioning in a well-known short story by Zora Neal Hurston, linking it to Gaston Bachelard's reinterpretation of Henri Bergson's theory of pure duration and self-creativity. In its treatment of time, the author and the story coincide. Hurston's work provoked serious criticism from the Harlem Renaissance movement as failing to be sufficiently sensitive to racial inequalities, an accusation Hurston countered by insisting that Blacks were not locked into a racially or class-determined past. Their lives were not merely reactions to history. Hurston's archaeological, autobiographical, and literary works all transcend ethnic and gender issues to take on larger human concerns. This particular story "Sweat" illustrates how the most impoverished individual is still able to initiate a deeply personal process of generating pure time, or duration, to effect two types of liberation: from her own passive resistance and from another's active aggression. In this way, Delia, the unlikely Black washerwoman heroine, moves herself out of her fated origins. The story's horrifying and ironic contrapasso justice confronts both the protagonist and the reader with a surprising but equally 'determined' conclusion as Delia is compelled to witness the violent death originally intended for her. It is a cliché to claim that Hurston's story enacts the moral that each ending is also a beginning: the battered wife's life begins as her abusive husband's ends. But McCloskey Engle's reading of the story radically subordinates this worn melodramatic panacea. Poetic justice, however well deserved, cannot rescue individuals from a destiny elsewhere ruled by race, class, or gender. Self-generated duration, the insistence on *one's own time*—in such quotidian activities as gardening, weekly worship, singing, pride in work, careful scheduling of chores—determines the future of a life.

Keeping with a panorama of violence in the form of twentieth-century terrorism, Marcus Bullock explores the theoretical options of comprehending the mechanisms of temporal imagination at work in terrorism, and ways to break the dominance terrorists gain over their prospective victims' outlook on the future. Bullock immediately sets the stage through the analogy of literary time, which is a coherence that shapes events into a form or plot that we recognize and that promises "more to come." The imaginative text of terrorism is precisely this—a hidden text that promises an impending return. Bullock succinctly concludes that the secret of terror is to dominate waiting. Referring to writers from Aristotle to Karl Kraus, Bullock situates the human capacity to redress this ubiquitous violence in poetic language, which deals with universals and keeps the convenient but corrosive division of 'them' and 'us' out of reach. Bullock forefronts Edgar Allan Poe's resolve that literature's absolute commitment to its own inner coherence, without regard for Duty or Truth, will fulfil homogeneous, empty time with time full of the now, or *Jetztzeit*. Here again, human creativity of a unique kind will recast 'waiting' into aesthetic time of a higher value. This inverted gesture of "Ursprung ist das Ziel" (origin is the goal) offers us access to a past that returns the present to a more rational future. The human capacity for poetic language reflects our resistant consciousness in confronting international terrorism.

The potent and dangerous symbols that serve to demonstrate Bullock's analysis of terrorism's discourse are equally prevalent in Masae Yuasa's forensic look at Japan's metonymic shaping of its own nuclear history.

Time imageries and their effects are central in her analysis of the conceptual framework at work in Japan's paradoxical commitment to nuclear energy, which, as she argues, originated in the experience of the atomic bomb. Yuasa introduces the theory of sociologist Yusuke Maki to explain standard patterns of imagining time, or variants of noo- and sociotemporal ordering, and illustrates how three different time imageries are involved in Japan's current debate on the use of nuclear energy. The dominant paradigm in support of nuclear energy is one of "segmented linear time," which envisions post-war Japan as moving from the catastrophic origin of the Hiroshima A-bomb, through the implementation of nuclear energy, to the goal of a technotopian future characterized by an abundance of energy and wealth. Yuasa finds a second paradigm in the inhabitants of the Iwai island, who are waging a fight against the construction of a nuclear power plant on their small island in the vicinity of Hiroshima. In her analysis, these people see their lives embedded in a temporal environment reflecting the cycles of nature. Their desire is not to progress toward a utopia of economic abundance, but to maintain the recurring cycle of life that has sustained them and their ancestors. Finally, she also remarks that large portions of Japanese society have ceased to believe in the dominant paradigm, and are instead informed by an image of time moving incessantly forward into a future that has no specific *telos*—an image that fosters, in her opinion, anxiety and an ego-centred focus on individual welfare. Yuasa thus wonders whether Japan is in need of a new version of segmented linear time, in other words, a re-interpretation of its origin and its *telos* as a nation to solve its present predicament.

Karmen MacKendrick ponders a different avenue of response to a society's traumatic experience. She carefully considers the ways that forgiveness re-opens the future in societies traumatized by violence and injustice. Her analysis begins by linking forgiveness to an originary creation in which time is for-given, in other words, in which time is given in advance. Creation stories derive their force from the fact that they cannot be entirely tied to justice alone since the dynamics of human potential would in this way be foreclosed. To escape the uni-direction of the first 'sin,' humans must negotiate ways to avoid the *telos* of their own sinfulness. Thus, forgiveness emerges as the creative human force distinct from divine retributive justice into areas of unpredictable openness. The past need not be repeated or retold; it need not, above all, be forgotten either. Forgiveness becomes a dimension of time—of the future itself. But miraculously, not only the future, but the past itself can be remade, can be altered. Where Hurston's protagonist immerses herself in 'pure time' in the present in order to redeem in advance a future otherwise closed and determined, MacKendrick's methodology of forgiveness witnesses victims and perpetrators meeting across the abyss of violence and death to regain a lost future for entire peoples and civilizations.

Understanding time is, as Fraser would often remark, an open-ended search because both its object and objective relentlessly evolve. This present volume represents, therefore, no more or less than yet another step on the continuing path of each time-questioning individual and the community of timesmiths that is ISST. It is our hope as editors that in reflecting on the concepts of origin and future, on the complexity of their mutual forces, and on various case-studies from literature and the sciences, this volume will deepen readers' understanding of what they encounter, whether they turn forward to the future, or back to beginnings.

Zurich and London, November 2012

SECTION ONE

ORIGINS EXPLAINED

FOUNDER'S LECTURE: THE ORIGIN OF THE INTEGRATED STUDY OF TIME*

J.T. Fraser †

I have often been asked why I have dedicated so much of my thinking to the creation of an integrated study of time. Perhaps a divine intervention? A single, memorable event? If neither, what, then? From the examined perspectives of five decades a clear answer to this question is now possible. It was, instead, the cumulative experience of growing up along the route that was taken by the Nazi forces marching to the east, the Russian army marching to the west and six centuries earlier, by the hordes of Genghis Khan attacking Europe. I found myself immersed in a vast historical drama, coming about and running its course in World War II in Europe.

When I entered the United States in September 1946—that is, soon after the end of World War II—I had a great deal of what came to be known as "post traumatic stress syndrome." The intensity and lastingness of my dedication to develop an integrated study of time has been driven—so I now believe—by an instinctive realization that dealing with time felt and time understood could help tame that stress and put it to useful work.

This paper is not yet another history of WWII but an account of a very few representative episodes of my war years. Together, they suggest my state of mind upon my arrival in the United States. It then turns to the birth of the integrated study of time and concludes with an assessment of its future. Therefore—to the story.

In Obuda in early 1944, my friend and fellow Boy Scout, Akela, and I were strolling and discussing girls as we walked by a large man sitting on a barrel. He said to himself and to us: "The tobacco flower girl is gone." He turned his thumb heavenward. Perhaps his lover, we thought. The world was at war: people did not discuss abstract things but whatever was important for them. This helped life go on. As did the fact that everything catered to the moment. Restaurants were open if they had food, coffee

* The manuscript was edited by Jane Fraser and Raji C. Steineck, making use of earlier versions and other papers by J.T. Fraser.

houses if they had roasted chicory roots brewed to a liquid that looked like coffee. Theaters and movies were open with censored shows.

It was not yet ten when I got home but Mama, Rosalie (our resident nursemaid and cook) and Grandpa Alexander were already in bed. It was an appropriate night for a young man to reflect and say goodnight.

Goodnight Mama. She was a daughter of La Belle Époque. The English called it the age of long peace. Women of her class enjoyed the accumulated treasures of their age and watched their prancing kings and princes with joy. Mama would often hum tunes from *The Merry Widow* for it reminded her of faraway Vienna, its music, its queen who died standing up sucking her thumb, of a prince who killed his lover, then committed suicide. Whenever she spoke of Vienna she looked like a disappointed angel because her fantasy did not match her reality. She was only seventeen when I was born, she told me, with the help of St. Anne.

Goodnight Grandpa Alexander. I loved and admired him. His ancestors placed their dinner between their saddles and their horses. After some time the rider's weight and the horse's heat made the meat just right. But that was over a thousand years ago. He was a country boy who worked the fields with those of his siblings who survived infancy.

Good night war. I was born five years after World War I officially ended in 1918. But the Versailles Treaty of 1919 left Pannonia sufficiently dismembered to make its citizens feel that, for them, the conflict never ended. In my child's logic, I looked for the causes of the ceaseless propaganda to hate this-or-that enemy. I found no causes but plenty of hateful signals. When I was still in elementary school, at a time described by Churchill as the gathering storm, opposite our apartment house was the toy store of Mr. Klein with Mr. Tapper in its window. A mechanical creature that nodded, turned left and right, tapped the window then looked at the kids with their noses pressed against the glass. During an October night someone was grabbed and used as a battering ram against Mr. Tapper's large glass pane. Next morning Mr. Tapper tapped on a bloody, broken window. The window was replaced but the blood reappeared; people said it was witchcraft. The store went out of business; the war went on.

Soon after the Goodnights, Grandpa Alexander died. Very early one morning thereafter, having had enough horrors, I was on the street watching silent men and women going to work. By noon the first German units arrived in motorcycles with sidecars. Later came the Wehrmacht in unending columns. On the same day a cousin of my father's showed up with his wife. He was one of the large store of cousins who belonged to the Jewish part of the family. The Pannonians had put them across the Polish border. They were told to run. They ran. But the Poles did not want them either. They drove them back to Pannonia and told them to run. They made it to Obuda. Homeless, penniless, dirty, and tired.

During the days that followed, groups of three men would appear at different places to pick up people whose names were on long lists. A civilian and two policemen, messengers of death, sent by Lucifer who remained out of reach. They would come between two and three in the morning when sleep was the deepest, resistance the least. A sinister fog descended upon the country, bringing forth what was worst in its citizens.

The lame woman doctor, who attended to my dying father, was accused of having performed an abortion. When two policemen and a civilian came to arrest her, she threw herself off her sixth floor porch. In my mind the story of Mr. Tapper, that of the man thrown against the window pane, and the suicide of the lame woman doctor were somehow connected to the beginning of Hitler's attack on Poland.

Unavoidably, I was called up for labor camp. I reported to the gathering place in front of the railroad station. Next to where I stood but cordoned off was a funnel-shaped area ending in a gate. Men, women and children were sitting on suitcases that held their earthly belongings. Just next to me: a middle-aged man, his wife, a daughter perhaps seven, a son maybe fourteen. The boy had a large, distorted head, tears freely flowing.

"She took it from me. She took my car from me." The distant eyes of the mother were without sparkle or tears. "Sharika, give it back to him. You know how it is." Mater Dolorosa, watching her son being crucified. The family began to walk toward the gate. Their bodies were pressed together tighter and tighter until they were murdered in the gas chambers of the East and burned to ashes in their crematoria. But all this I did not reconstruct until many years after I watched the sad family depart.

The labor camp was in northern Pannonia. It soon became an unguarded prison. The German colossus had begun to fall, and Pannonia hovered at the edge of total chaos. After a few weeks, some of us decided that it was time to escape. Four men, a row of organ pipes. Four shadows very early on an October morning. We walked inside a barbed wire fence, then through an unattended gate. With me walked Bert. A janitor. He carried six cans of sardines. He loved folk dancing. It was easy to see that he had no family: his face did not darken at unpredictable instants as did the faces of men who did have families. Mat the lawyer walked next. He spoke of his mistresses and bragged that he defended socialists in court with sufficient success to get himself deported. Dr. Munk, a former academic philosopher, was the third among us. I was the youngest in the line, with no record except that I did not fit any established category of the enemies of the state. As the last organ pipe, I thought of my boy scouting years with sunshine and happy songs.

The scene changes as the four organ pipes hunted about for someone to hide them. When we arrived at the cabin of a forester, he agreed to put us up in return for our promise that once the Russians arrive we'll put in a good word for him and his village. How ignorant we were.

It was dark and snowy. Around noon a dirty and jittery soldier showed up, after an hour or so he left without a word. Two days later snow flurries lasted into the night. "It's unreal," I said. "Not if you hear what I do," said Munk. A distant hum was heard from the south. Later on, low aircraft. Russian or German because the Americans and the Brits flew very high. I expected a ghost to appear and it did. It began as a whitish dot on the window, turned into a nose then into a head with a cocked hat. The straggler came in, put down his rifle and went to sleep. In the morning he told us about himself; we told him about ourselves. We believed nothing he said; he believed nothing we said. We parted in peace and mutual distrust.

Evenings found me leaning against a fir tree watching the fires along the horizon kept ablaze by a furious divinity. They were the muzzle fires of Stalin organs. It was impossible to believe that there might have been places on earth less insane than here. One evening a woman walked in, warmed water in the basin, poured Borax into it and sat down to soak her feet.

A few days later came a platoon of Germans: helmets on their knapsacks, machine pistols in hand, canteens on their belts. No one in the hut was supposed to have understood German—Pannonian peasants are we all—I was pushed forward to do the theatrics. The Germans pulled down the curtain, revealing ancient Anna holding the women; the kids holding onto whatever skirts they could. Two of the soldiers grabbed Dr. Munk. "I told you so," said Munk. He was hit on the shin with a rifle butt. As he backed out the door his face showed satisfaction rather than fear.

One of the men in the hut wanted to know, "Where did the gentleman go?" "He did not go. He was taken." Was he really that stupid or that glad, to see a member of the intelligentsia dragged to execution? Nothing was said thereafter about the incident. As I lay down on an old rag, I thought of Roosevelt talking into many microphones.

A few days later three shadows were leaving an ugly present for the other side of darkness. The women and kids lined up in front of the hut. One of the girls waved an embroidered handkerchief. We picked up other men—refugees, prisoners, as you wish. By the time there were a dozen of us lost walkers, we were joined by Russian soldiers who rushed us to another village with wall-to-wall soldiers. There, the houses were charred, windowless, but still had walls. A soldier approached, poking his bayonet into a knapsack. Bert must have felt he wanted to reach out to another human; he began to recite a Russian sentence that he had memorized. It was supposed to have meant, "We are your friends and enemies of the Germans!" Hearing this, the soldiers started shouting *Spion! Spion!*

Next morning we had a talking to—in German. "You are now free, free of the Nazi yoke and Hitler is dead. You are privileged because you will help rebuild Russia and atone for your crimes. Then you may go home and eat as much as you want!"

One evening we slept in a room with its walls covered by bookshelves. Someone shouted, "Books! Now we can cover up the shit in the back!" Pages were torn out. "Don't be selfish," said a voice, "leave pages for other asses!" I stopped and held one of the books that passed me. A peek into heaven. *Aphrodite* by Praxiteles. The Caryatides. Donatello. A park in Rome. I sent it on fast. I feared that my fantasies could drive me insane. A day or two later I saw houses with people in them. Women of uncertain ages peering out the glassless windows. One was nursing a child perhaps three years old. A woman wore her wedding gown. By local custom, widows were buried in them. That day I realized that all this misery-in- motion was on Moscow time. We were awakened at two A.M. local time; chased on to other rags of lice, for sleeping, early in the afternoon.

One morning an angel from heaven came. A young Russian in uniform. He stopped in front of me. "I am an architect specialist. Come with me," he said in German. He took me to a room where two Russian women operated a transmit-receive radio (made in Cincinnati). And two Italians cranked a generator. Benedetto and Ugolino. Ugolino spoke. "We'll never get out of this hell hole but you will. I'll steal some paper and a pencil. I'll give you our home addresses. After you get out, write our wives. Tell them everything." I hid the addresses in the sole of my boots. I began to scout around for others who disliked it here—and would make good companions. I found a Turkish lawyer, a Tito Partisan and—yes—an English soldier.

He was John Gordon Tasker of Hull. He fought in Greece, was captured by the Germans, transported, like cattle, in a freight car. "Beyond Belgrade the partisans blew up a bridge; the German guards were taken care of. I was hit on my neck. When I came to I was in the house of a peasant family. As the Russians came closer I walked to meet them. I gave up. They stole all my documents. These people have no manners."

On New Year's Eve I walked to the garbage dump because the word was that one could find pieces of not-yet-rotten potato. When I got back to the room, John said it was time to sing "Auld Lang Syne." "People hold hands like this." We all knew the tune; he recited the words.

> Should auld acquaintance be forgot,
> And never brought to mind?

"Robby Burns, you know, but the tune is older." We hung together by an inner perception of what was noble and worth preserving. Far above us Robby Burns, late of Dumfries, Scotland and Ludwig van Beethoven, late of Heiligenstadt, Germany celebrated a small island of civilization on the dark plains of Pannonia.

> Freude schöner Götterfunken,
> Tochter aus Elysium...

A few days thereafter a handful of Ukrainians with belts in their fists and hatred in their eyes pushed some of us onto a railroad platform and into a car of a long freight train. Barbed wire around and between each car. Dirty polar bears with shiny red stars riding shotgun. In the car I found myself next to one of the sliding doors where there was a four inch square hole with layers of shit and urine frozen on it. Once a day, three Russians opened the door opposite the hole, twisted their noses and left a pail of thin soup. Rumors began to circulate. Two people had typhoid fever. Then—whenever—the typhoid people died and went out when the soup came in. "*How unprofitable seemed to me all the uses of this world.*"

A week or two or an eternity later, a young man gave us the weekly homily. He reminded us that the rich, fat people of America and England admire the Russians because they are winning the war against the Germans and that Hitler was dead. Some time later—count lost—the train stopped near a band of Pannonian gypsies who kept on playing the Communist International with abandon, again and again, making that stirring march sound like a Vienna waltz.

> This is the final conflict,
> Let each stand his place;
> The International Party
> Shall be the human race.

I espoused the outrage but rejected the solution. Specifically, I volunteered for *robota*. Work. Out we marched. A sullen old man, two old women, two boys of my age and I. All but myself, Romanians.

We kept on walking. Later on, the distance between us, three young men pulling the cart and the three older people began to grow. First at a painfully slow rate, but later it increased as the two boys and I took to our heels, rushing the cart. Our rush stopped when we arrived at a country cottage and—miracle—sat down for bread, beans and cheese. An old man said the Greek Orthodox blessing, "Christ has resurrected." The answer we gave came in Romanian. 'Indeed!"

Three of us, young men, slept in a loft short of hay but long on heaven. Breakfast was bread and wine, siphoned from a barrel with a thief tube stuffed up with gunk. The two older women reappeared. "Boil water, pour in, snap with knife," said I. One of them gave me a kiss and a portion of *mammaliga*. At dinnertime the old man looked at me and took stock.

"We are all counted here, as is the food. You'll have to leave with the wood." So it was that the next morning I took off for Bucharest on a rickety wagon, carrying wood, to search for an aunt I never had, but whom I placed on Callea Mossilor because I was told that it was the fanciest part of town. It was late afternoon when I asked to be let down. I was in a workers' district. Surely the best part of town for a tired and hungry young man who had no place to sleep, no money, no food. Should I keep on walking until I collapse? Knock on a door?

A man and two women came by. He wore sandals and a pair of puttees, the women wore what looked like prison garb. Pannonian was never a popular language here and I did not know Romanian; I tried my poor French. "Liberatee de *Lager*. Je suis solitaire." "Komm mit" said the man in accented German.

Eventually, we stopped in front of a house. "You'll be staying with Madame Fanchi" said the man, "but will have to report to the Center once a week. *Shalom*." I did not know what "Shalom" meant. I took it to be a word with which Hebrew-speaking people greet each other.

Fanchi had two sons. That evening and many days and nights thereafter, I played dice and cards with the Fanchi children while their mother's friends came and left. The kids explained that all these men were their uncles. Some of them stopped to join our games. As I came to know Fanchi, I found in her a great deal of goodwill as well as pettiness. She ruled over her kids with gentle tyranny. Food was in very short supply, so rather than eat she said, "I am not hungry" or "You don't have to finish everything on your plate." After some days at that secular heaven all I had left from those horrid months were the papers with the addresses of Benito and Ugolino and that of John Gordon Tasker in Hull.

At the Center thick rumors floated around. In the Urals children of twelve were mining something called Uranium. In Poland there was a village where the Germans built shower rooms for the purposes of gassing prisoners. Family members held onto each other so strongly that after they died, their bodies had to be cut apart with axes. All this was a part of Hitler's Final Solution.

I have to cut short the details of the next months except that, I did find the British representative in Bucharest and delivered John's letter. In response, they helped me to get to and on a train leaving for Obuda to try to locate my mother. From there to—my choice. All this in return for taking a letter to someone in Obuda who was working within their *aliyah bet*. That is Hebrew for an underground conduit to the Holy Land, into which I planned to insert myself until I located my mother.

"That's a deal," I said to my contact, "Shalom."

I did find my mother. I was able to bring her to the USA after Stalin died. She herself naturalized and died in this country and is buried in Pleasantville, New York.

There and then, in Obuda, I had to complete what I had promised to my contact with the illegal immigrants to Palestine. Therefore, on an appointed day I appeared at an agreed upon spot. A short, wiry man welcomed our small, strange brigade. "We are young workers going to work on a large farm," telling us who we were. "We should sing." We did. We passed many hamlets but there was no barking. All dogs have been shot. The stars in the east began to fade. In the west they remained brilliant, beckoning to the homeless.

Our education continued. When the Russians reached the Austrian border—we learned—they started building a no-man's land, a fortress wall of sorts.

"Shalom" said the wiry man. "You are now in the West."

Many things happened between then and when I entered the USA as a student of Columbia University. It included the tail end of my homeless years, and being declared a Displaced Person by the powers that be.

Fortunately, I was able to enlist the good offices of Professor Charles Rufus Morey of Princeton, then American Cultural Attaché to Italy, who guided me to the Allied Control Commission. There I was hired to work on the immense amount of inquiries from people looking for surviving family members and friends. He helped me gain admission to Columbia, obtain a student visa and sail to the US on SS Marine Shark—a troop ship carrying returning servicemen and women.

I settled in Manhattan, doing dishwashing and selling hot dogs and hamburgers at Grand Central Station. I was then hired to work as a cataloger for Columbia's Butler Library. Later, I was naturalized by a special Act of Congress through a bill introduced by the late Senator Flanders of Vermont. But the many details would delay my report about turning to—time.

Thence, to Manhattan. I was walking along the Hudson River, searching for ways to order my memories and feelings about the churning details of my life. What general principles might I employ to order, sort, and assess the past? I had found myself between two multitudes armed to the teeth. I knew that I was watching a struggle between two ideologies, each of which was convinced that it, and it alone, was destined to fight and win the final conflict of history. Having been aware of both dogmas, I came to wonder whether there does exist such a thing as "a final conflict."

My preoccupation guided me to write a senior undergraduate paper called "A Short Essay on Time." It earned a National Engineering Honor Society award for "excellence in the humanities by an engineering student" (1950). A decade later I came upon two seminal works that taught me much and encouraged me. They were G.J. Whitrow's *The Natural Philosophy of Time* (1961) and S.G.F. Brandon's *History, Time and Deity* (1965). My efforts continued in *The Voices of Time—A Cooperative Survey of Man's Views of Time by the Sciences and by the Humanities* (1966) and by the founding of ISST (1966).

Watching the clash of cultures and the attendant release of primeval emotions stripped of the usual niceties, I had to conclude that man is only superficially a reasoning animal. Basically, he is a desiring, suffering, death-conscious and hence, a time conscious creature. I realized with a pleasant shock that the question whether there can exist a final conflict was too crude to be fundamental. Namely, it is possible to imagine a world without mass murders, but it is not possible to imagine humans who will not declare, in innumerable ways "Death, be not proud..." because the conflict that gives rise to such a rhetorical command—the conflict between the knowledge of an end of the self and the desire to negate that knowledge—is at the very foundation of being human.

In due course these ideas grew into a full-fledged theory of time as a nested hierarchy of creative conflicts. Its main propositions are as follows:

1. Nature comprises a number of integrative levels
2. which form a hierarchically nested and evolutionarily open system

3. along a scale of increasing complexity.
4. Processes characteristic of each of these levels
5. function with different types of causation
6. and must be described in different languages.
7. Each level determines a qualitatively different temporality, and
8. each adds new, unresolvable conflicts to those of the level or levels beneath it.

How may one sum up an open-ended report like this? Two poets will help.

About the conditions that gave rise to the idea in general, listen to *The Rime of the Ancient Mariner* by Samuel Taylor Coleridge (1772–1834)

> Since then, at an uncertain hour,
> That agony returns;
> And till my ghastly tale is told,
> This heart within me burns.
>
> I pass, like night, from land to land;
> I have strange power of speech;
> That moment that his face I see,
> I know the man that must hear me:
> To him my tale I teach.

As to the future: here is advice from a Welsh bard, Dylan Thomas,

> Do not go gentle into that good night.
> Rage, rage against the dying of the light.

THE SEARCH FOR ORIGINS AS DETECTIVE STORY: DETECTING DATELINES AND THE MYSTERY OF ORIGINS

Paul A. Harris

Abstract: The essay compares two modes of investigating temporal origins: a scientific detection of dates of origin, associated with the chronometric revolution across a number of fields, and a speculative investigation into the mystery of origins in a more absolute sense, where the physical and metaphysical shade into one another. The former mode operates according to a linear, causal temporal model, while the latter initiates an iterative, recursive series of investigations from present through the origin and back again. A range of examples from science, theology, philosophy and literature are discussed.

Keywords: Origin, mystery, chronometric revolution, Cretaceous-Tertiary K-T Extinction, dendrochronology, palindrome, vortex, J.T. Fraser.

Introduction

The theme of the 2010 triennial conference was "Origins and Futures," a phrase which immediately evokes some sort of open-ended arrow of time, anchored by a beginning point or points ("origins"), and pointing towards indeterminate pathways or trajectories ("futures"). The pregnant phrase "Origins and Futures" implicitly situates 'us,' the time-conscious subject, at a complex cusp articulated as the connective 'and'—to be a noetic (hence temporal) human being is to find oneself between origins and futures, in the knotty space of mapping oneself into a time line extending from origin(s) into future(s). Scrutiny of such a linear arrow of time soon leads to complications, however—retrospective memory and anticipatory imagination project multiple lines of flight that only a multi-dimensional arrow could indicate. Fortunately, Rudy Hilt has provided precisely such an image with "The Arrow of Time," which graced the cover of our conference booklet.

Convention decrees that the President's Address cover the conference theme in some sort of comprehensive overview. Since "origins and futures" encompasses a rather wide swath of time, it seemed only fair, not to mention pragmatic, to truncate the topic by bifurcation. This essay thus addresses only one aspect of our theme, namely origins, and in turn divides investigations of origins into two different kinds, which I'm calling "detecting datelines" and "the mystery of origins."

Detecting datelines is a mode of looking into the past that seeks to pin down a point of origin, or ascribe a temporal co-ordinate to an event or object. This mode of investigation may be conceived as a form of detective story, in that its governing question is "what happened?" or "how and when did it get there?" and the goal is to reconstruct a past event and date it as precisely as possible. This reconstructive process essentially entails a form of detection: signs in the present are interpreted as traces of past events. The 'detective' undertakes an empirical investigation in which the object being scrutinized serves as the equivalent of a crime in detective fiction: it is the origin or cause of the narrative, and the narrative's plot depicts the process of reconstructing what happened. In a sense then, the detective plot involves the reconstruction of its own origin (see Brook).

The mystery of origins encapsulates a mode of looking into the past that contemplates and speculates about some origin of an absolute kind. This mode of investigation may be characterized as a form of mystery story, where the governing question is "what is an origin?" or "how did this origin itself originate?" These questions afford only solutions that themselves remain irreducibly mysterious, at least when framed in metaphysical terms. Consequently, the resolution of the mystery of origins involves moving around, in and through an origin without solving or resolving its mystery. Some literary critics point out that the mystery genre often touches on or broaches supernatural or transcendent domains because the investigation of a mystery organically expands into contemplating *the* mystery—namely, the mystery of whether or not there is anything beyond the material. Mystery denotes something secret, obscure, or puzzling; it poses as an unknowable truth or Gordian knot. The word's etymology conjures clouded eyes and lips; connotes a theological secret; connects to a desire to initiate or teach. In a mystery story, then, the investigation of the mystery returns to its origin, as in detective fiction. However, it becomes a mystery when the origin or truth proves undetectable. Thus no final solution can be reached; investigation must continue returning to the origin but without expectation of an endpoint (see Irwin).

Each of these investigative modes is governed by and constructs a different mode of temporality. Detecting datelines operates within a linear, causal model of time, in which solutions result in plotting the points and connective features of events. This is the narrative mode favored by science in particular. Mysteries of origins unfold or disclose a recursive, tangled, cyclical or vortical model of time, in which solutions become themselves the subsequent mystery to be investigated. This is a narrative mode peculiar to poets and kindred spirits across the disciplinary spectrum.

Detecting Datelines

As human curiosity and ingenuity continue to create more and more sophisticated prostheses for our senses, the parameters of our investigations of the past increase exponentially. And as our instruments allow us to peer into ever deeper periods of past, those periods themselves recede further and further from the present. As recently as the 18th century, most Europeans believed that the universe (which they would have called the world or creation) was about 6,000 years old. Today, billions of humans believe that the universe is about 13.7 billion years old. Equally impressive or even astonishing is that as we look further and further into the past, our ability to date things also grows more and more precise. Historians point to a 'chronometric revolution' that has been occurring since World War II: a series of technological innovations in measuring and dating objects and events (see Hedman). These techniques include radiometric techniques, carbon dating, genetic techniques and DNA dating, the study of plate tectonics, and astronomy. These innovations change our sense of what history itself includes: historians traditionally define history as written history. The chronometric revolution allows historians to study 'Big History': a history whose sweep encompasses the history of the cosmos, terrestrial and evolutionary history, as well as human cultural history (see Christian).

Practices of detecting datelines differ widely by discipline, of course. But in many fields, new chronometric instruments and techniques have enabled people to pursue intriguing investigations of longstanding mysteries, and produce persuasive, provocative solutions to them. These investigations often take the form of detective stories in which a satisfactory solution to some unexplained event or unanswered question in a field's historical record is found at last. Such enigmas are events or effects whose causes remain hidden from view because we have not uncovered or been able to make visible the appropriate evidence. Often this is a function of the limits of clues or evidence: the incompleteness of the fossil record, etc.

In the paleontological record, one such mystery concerns what is called the Cretaceous-Tertiary K-T Extinction. Sixty-five million years ago, 85%

of living species died within a short time. This is popularly known as the question, "what killed the dinosaurs?" Walter Alvarez, the scientist primarily responsible for discovering the now accepted solution to the mystery—it was caused by the impact of a huge meteor hitting the earth—wrote a marvelous popular account called *T.Rex and the Crater of Doom*. Alvarez and his father Luis (also a Nobel prize winner) discovered in geological investigations that the element iridium was often found in the strata at K-T boundary. This led them to postulate a hypothesis that a meteor must have struck the earth around this time. The hypothesis found confirmation with the discovery in 1981 of the Chicxulub crater.

Different means of detecting datelines produce results that can be quite stunning. A compelling case that comes to mind involves what might be called a real-life "Art History Mystery," to play off the emerging popular genre. Like many fictional art history mysteries, this case revolved around dating an object to ensure its authenticity. In 1971, J. Paul Getty bit the bullet and bought a large cabinet purportedly made in the Renaissance in France. In doing so, Getty bucked the advice of his curators, who did not believe the piece was genuine. Their investigation found that the cabinet was in suspiciously pristine condition; moreover, the surface was coated with colored wax, suggesting that someone had tried to make it look older than it really was. These experts concluded that the cabinet was a product of the Renaissance Revival of the 19th century, when American industrial magnates snapped up Renaissance-style furniture, including many fakes, from cash-strapped European aristocrats. Because of the doubts about its authenticity, Getty only paid $1,700 for the cabinet—a far cry from the asking price of $46,640 some 50 years before. The dubious cabinet was never displayed in the Getty Museum—until it became a fabulous story of detecting datelines, a story told in its own exhibition. The detecting of the cabinet's dateline combined many chronometric techniques, but was clinched by path-breaking work in dendrochronology, the study of tree rings as a means to measure time.

Dendrochronologists look at the pattern of wide and narrow growth rings in a piece of wood and compare these rings to weather patterns in different regions over time. Wide rings grow in wet years, narrow rings in dry years. Scientists can date a piece of wood by comparing its tree-ring patterns to the tree-ring patterns of other pieces of wood felled in different places and times. This was one prong of the investigation of Getty's cabinet. Scientists examined the ring patterns in 17 pieces of the cabinet's structural panels, which are made of oak. This is of course a delicate process when the wood in question may be very old. Where the end grain of the panels was visible, they took high magnification photographs. Where the end grain was not visible, they made high-resolution X-rays.

Dendrochronologist Didier Pousset then used the photographs and X-rays to create an average growth chronology of the oak in the cabinet. For a dendrochronologist, an average growth chronology is like the fingerprint of a tree. Pousset then compared the fingerprint to several master chronologies, records of woods whose ages and origins are known. Each master chronology is created by aligning the tree-ring series of living trees and the tree-ring series of documented architectural timbers, such as from well-known buildings. By overlapping the rings of increasingly older wood, a virtual 'endless tree' can be constructed that extends far into the past. The analysis revealed that the oak used to make the cabinet came from a tree that was cut down in the autumn or winter of 1574–75 in the region of Burgundy, France. Since the wood would have had to dry (or season) for several years and the construction of the cabinet would have taken some time, this analysis strongly supports the date of 1580 painted on the cabinet.

The Mystery of Origins

It seems tautological to observe that origins mark beginnings. But when we try to ferret them out, detect them, reconstruct them as any kind of 'point' or originary moment as such, origins remain shrouded behind a veil. As Philip Kuberski succinctly expresses the matter, "Origin gave way to the world we see around us, yet it keeps to itself, far from us, withdrawn in time and space–nothing more than an opening now enigmatically closed" (434). To evoke or speak of origins in relation to time, as our conference theme does, also would seem to border on the tautological, since origin is an inherently temporal concept and origins are inherently temporal events. Finally, in yoking time and origins together, our conference sounds the ultimate mystery: the origin of time, which would serve as the origin of all origins.

Under the sway of contemporary cosmology, to contemplate the origin of time is to consider how the universe came into being. Accounts of the origin of the universe (or "creation," depending on one's approach to the issue) comprise a genre that includes creation myths, scientific theories, and religious texts. With his usual dry wisdom, J.T. Fraser summarily states: "The earliest stories of cosmic birth are set in the countryside where the narratives originated and report about the conflicts among gods, between people and their gods, and among people. The later history of cosmogonies and cosmologies shows an expansion of spatial and temporal horizons as well as changes in the preferred modes of perceiving and explaining the nature of the world" (67). The primary of the preferred modes Fraser refers to is of course mathematical cosmology, driven and supported by astronomical observations that look further and further into the past. If cosmological tools have enabled scientists to treat the creation of the universe as a kind of detective investigation, however, this investigation does not ever provide an answer to the mystery of the origin itself.

On one level, cosmology is a game of detection that involves finding and interpreting clues composed of light. These clues are traces from the past: as Eric Chaisson puts it, "Astronomers are the ultimate historians; our telescopes, effectively time machines." The "cosmologists' dictum" is that "*looking out in space is equivalent to probing back in time*" (7). Yet no matter how powerful a telescope we ever devise, we will never see anything earlier than the first stars forming about 200 million years after the Big Bang, for two reasons. First, until about 380,000 years after the Big Bang, the universe would have been opaque. It would have a photosphere, the term for density and temperature of gas that make it become opaque because no photon can escape. Secondly, the glow left from the Big Bang is only visible at a red shift so great that the light is shifted out of the visible to the radio range, and so we detect it only through the cosmic microwave background. The origin of the universe, then, will remain shrouded in the darkness before "first light," as cosmologists call the first moment when a photon can escape a dense gas (see Kidger 178–179).

While observations of the origin of the universe thus prove impossible, descriptions of the origin can be formulated using symmetry and scaling arguments: scaling models back in time leads to a description of a cosmic singularity at the moment of the Big Bang. The Big Bang is not itself an origin though—it is rather a map of the immediate aftermath of the singularity at time zero. The singularity itself is a construct: its initial or boundary conditions are inserted manually into the model.[1] And to investigate how the origin came to be, to inquire after its cause, is to venture beyond the reach of science. Most cosmologists share Chaisson's agnostic stance that "queries about the nature of whatever existed before the bang

[1] See Heller (ch. 3) for a clear treatment of this problem.

amount to inquiries less about the origin of the Universe and more about the origin of the origin," because, quite simply, "there are no data" for what happened before, rendering it not "experimentally or observationally testable," and therefore not science (57).

For the cosmologist, then, the beginning of time is undetectable. But even being able to have data or observations of the Big Bang or conditions similar to it would not probe the further mystery that underlies and comes prior to the universe's physical beginning, namely the origin of the universe. This origin is particularly mysterious because it stands for the ur-origin, the origin of everything and hence all other subsequent origins of other kinds. Two courses seem possible at this point. One way involves embracing an eternal, absolute cause, principle or deity. Augustine provides the most renowned version of this solution to the mystery. For Augustine, the origin of time is a corollary of the origin of creation. Eternal God created time and the universe or world together: "the world and time had one beginning and one did not anticipate the other" (*City of God* 350). What was God doing before He created the world? Augustine's first answer is refreshingly humorous: a specific circle of Hell has been designed for those who ask such questions. His ultimate answer, if less enjoyable, appears logically sound: since God is Eternal, while time and the world are by definition temporal, then it makes no sense to ask what God was doing 'before' the creation because such a question presumes that time pre-existed time and the world (*Confessions*; ch. XIII). Fraser makes the interesting argument that by time Augustine would have meant not a time of the cosmos or a time of physics, but "noetic time," the human experience of the passage of time. In other words, Augustine only seems to speculate about the origin of time—he in fact is contemplating the origin of time as humans experience it.

Alongside Augustine, one finds many more mystical modes of grappling with the mystery of origins. Another way is to follow the Way—the Daoist Way, that is, of Chuang Tzu:

> There was a beginning. There was a beginning before that beginning. There was a beginning previous to that beginning before there was the beginning.
>
> There was existence; there had been no existence. There was no existence before the beginning of that no existence. There was no existence previous to the no existence before there was the beginning of the no existence. If suddenly there was nonexistence, we do not know whether it was really anything existing, or really not existing. Now I have said what I have said, but I do not know whether what I have said be really anything to the point or not. (5)

The attempt to find an origin point ends in the ground under one's feet becoming unstable and the line between existence and nonexistence becoming effaced or suspended. And perhaps, on some level, that is precisely the nature of origins: to occupy the zone where existence and nonexistence become indiscernible.

The origin of time poses a conceptual paradox in exactly these terms—the origin must be both nonexistence and existence. Put in chronometric terms, the origin of time, t zero, is imagined as a dimensionless point on a timeline. On the one hand, this point must mark a discontinuity: as a beginning, it must be qualitatively different than what preceded it. Yet on the other hand, t zero is a continuity: it is the first moment in the temporal sequence. T zero is therefore both outside and inside of time: it marks a paradoxical cusp where non-time turns into time. T zero would be the chronometric map of the Augustinian creation *ex nihilo*: t zero as both nothing, sheer absence, and a transition from nothing into presence, the emergence of the world from nothing.

Intuitively, the *ex nihilo* model of origin entails a pure being at the ground of time, which then starts moving—out of stillness, change and becoming. However, Fraser reminds us that cosmology implies the exact opposite picture: the cosmic singularity as sheer chaos forces us to see that: "It was being that arose from pure becoming, not the other way around. The origin of time is identically equivalent to the emergence of the most primitive form of orderability, the first break in the symmetry of primeval chaos, so called because of its homogeneous lawlessness—in which space, electromagnetism and gravity were all merged into a furious foam" (78). The origin as pure becoming has been imagined in various forms, including a quantum vacuum in which forms and matter appear and disappear at infinite speed.

The mystery of origins, then, ultimately unfolds as a quest or investigation without end or solution. The mystery of the origin impels us to move retroactively back along a timeline, make a pass at or through a ghostly origin, and return to the present to report on the non-findings of the journey. This investigative process creates its own temporal dynamic though: it essentially bends linear time into a cycle that is doomed to be reiterated. The origin must be a self-generative source of time, and the process of discovering and defining the origin becomes entangled with itself: characterized by the conundrums of self-reference, repetition, reiteration, or circularity. The mystical figures of time as cyclical origin include the Ouroboros.

As a professor of letters, I would like to evoke palindromes: palindromes are the linguistic figure that mirrors the mystical bending of linear time back on itself into a repeating or recursive spiral in language. In a palindrome, the linear movement of language reverses itself at the end, bouncing back and forth indefinitely. In this dynamic conception of the palindrome, the string of letters are imagined as mirroring one another across a pivotal point. This point would be the 'origin' of a palindrome, if we imagine the palindrome as signifiers in a linear sequence; the origin would serve as the zero point on the string of letters, with mirroring letters resembling positive and negative integers on a number line. Interestingly, the pivot of a palindrome can be either a letter or the interstitial space between letters, often in a single word—the pivot can be absence or presence, and thus conceptually functions as a place where presence and absence become indiscernible. The links between the palindrome form, and the ambiguous co-persistence of presence and absence around the origin, are expressed in the mystical writer Edmond Jabès's title *Nul L'un*: the ineffable origin (null) of the One, the name of God.

Let me conclude by discussing a particularly rich attempt to express the mystery of origins, in which several strands of this discussion converge. This work describes the creation of the universe through a poetic genre that enables language to become a kind of abstract diagram—it offers both scientific description and poetic-spiritual invocation of the origin of creation. The work is by Ronald Johnson; it is a concrete poem that ultimately constitutes a quasi quantum vacuum composed of elemental letters arranged in a periodic chain.

CHRYSOS
anthemon

f lux f lux f lux f
lux f lux f lux f l
ux f lux f lux f lu
x f lux f lux f lux
f lux f lux f lux f
(Beam 12)

This poem celebrates the birth of the universe as a flood of light. Johnson splits the word flux into composite parts, f and lux. As a discrete signifier, lux denotes a scientific unit of luminous emittance. The f in "f lux" may be read as a musical signifier, *forte*; the concrete poem thus embodies a loud emittance of light. Enunciated phonetically, the poem reads "EFF LUX," efflux, meaning an outflowing of light, connoting the efflux of God. Finally, f lux may be read as an abbreviation for *fiat lux*, the genesis of creation in *Genesis*, "let there be light." This concrete poem is a creation story, where religion and science merge in a vision of creation as a transcendental light shining forth. The poem's enigmatic title tells us that the light it emits is pure gold; CHRYSOS is Greek for gold. "Anthemon" is the Latinate form of chrysanthemum, the gold flower; CHRYSOS plus anthemon gives us chrysanthemum. The flower image places us in a garden of divine creation. The f lux poem is, within ARK, part of the "Foundation" section made of poems called "BEAMs"; it is thus composed of golden light, a divine light emitted by the sun. Simultaneously, this creation story also contains the birth of human life and its mortality: efflux also means expiration. The poem thus celebrates flux as becoming, change, and movement on the one hand, and on the other hand, it contains the fall from the divine into human mortality, expressed in efflux as expiration. In addition, a derivative of lux is Lucifer, the angel who fell away from God. In the cosmos, a golden garden of divine light, humans are fallen angels.

This creation story also may be read as a scientific description of the birth of the universe. This origin is an event in which an initial wholeness is divided into parts that display both particulate patterns and dynamic design. The particle-letters (f, l, x, lu, ux, etc.) may be aligned into patterns along several axes—horizontal, vertical, and diagonal. The universe is a function (f) of light (lux). Simultaneously, the letters move in a dance of motion that circles on itself; its end reiterates its beginning; the single bloc implies infinite iteration. The bloc of letters swirls in a dynamic flux; they form a vortex. Flux as flood, a continuous flow, is conjoined with flux as continuous movement or change, meaning difference. The poem's dynamic movement thus infuses a fusion of discrete elements; another meaning of the word flux is a substance used to promote fusion. Thus the word flux broken into pieces itself expresses the binding together of what has been rent asunder. Johnson recalls a trope used by Lucretius in *De Rerum Natura*, one inspiration for *ARK*: the trope of letters as atoms, and writing composing the world anew. Johnson's concrete poem here, in its dual nature as discrete elements and continuous flow, could be seen as a diagram that maps the poetics and physics of Lucretius. Letters function as atoms, elemental building blocks, swirling in a vortex. The poem could be read as a metaphoric echo of the periodic table, where letters signify elements that combine into molecular matter. The poem's interplay between the discrete and the continuous also captures the double function of the clinamen within Epicurean physics. As Michel Serres observes in his analysis of Lucretius, the clinamen charts both a solid-state mechanics of matter, and a fluid mechanics of turbulence. As solid-state mechanics, the clinamen is digital or discrete; as Serres says, the clinamen is "thus a differential, and properly, a fluxion" (4). As the swerve that initiates a fluid mechanics, the clinamen is analog or continuous, marking a fluctuation. The vortex is the site where disorder and order interact along these two lines: "The vortex is thus the pre-order of things, their nature, in the sense of their nativity. Order upon disorder...: the vortex arises by a *fluxion* in the first hypothesis... and by a *fluctuation* in the second..." (*Birth*, 31). Johnson's concrete poem, his flux-vortex, ultimately functions as a kind of matrix, in different senses. Literally, the poem resembles a mathematical matrix, an array of discrete elements that combine in different ways. Metaphorically, the poem is a matrix in the sense of a place of origin, or more fundamentally, a womb. The flux poem may be read as a diagram of two different kinds of primordial creative events: as a cosmic-religious description of the origin of the universe in terms of light, and as an atomistic-physical description of the origins of order and disorder in terms of the clinamen.

Perhaps every attempt to journey to an origin is really a construction of a matrix, a womb that could contain the origin in a protective envelopment. It is difficult to imagine time's origin as anything but a womb of everything rather than nothing. In moving from back-tracking the linear history of things to detect datelines to swirling into the vortex of origins, we end up making up a matrix of our own, assembling pieces that can be combined in different ways in order to evoke what ultimately remains a wordless silence.

References Cited

A Renaissance Cabinet Rediscovered. Exhibition November 22–August 5 2007. Los Angeles: Getty Museum. ‹http://www.getty.edu/art/exhibitions/cabinet›.

Alvarez, Walter. *T.Rex and the Crater of Doom*. New York: Vintage, 1998.

Augustine. *City of God*. Trans. Marcus Dodd. New York: Modern Library, 1950.

Augustine. *Confessions*. Trans. R.S. Pine-Coffin. New York: Penguin Books, 1961.

Brook, Peter. *Reading for the Plot*. New York: A.A. Knopf, 1984.

Chaisson, Eric. *Epic of Evolution*. New York: Columbia University Press, 2005.

Christian, David. *Maps of Time: An Introduction to Big History*. Berkeley: University of California Press, 2004.

Chuang Tzu. *Basic Writings*. Trans. Burton Watson. New York: Columbia University Press, 1964.

Fraser, J.T. "The Secular Mystery of the First Day." In *Time and Time Again*, J.T. Fraser, 67–80. Leiden: Brill, 2007.
Hedman, Matthew. *The Age of Everything: How Science Explores the Past*. Chicago: University of Chicago Press, 2007.
Heller, Michael. *Ultimate Explanations of the Universe*. New York: Springer, 2009.
Irwin, John T. *The Mystery to a Solution*. Baltimore: Johns Hopkins University Press, 1994.
Johnson, Ronald. *ARK*. Albuquerque: Living Batch Press, 1996.
Kidger, Mark. *Cosmological Enigmas*. Baltimore: Johns Hopkins University Press, 2007.
Kuberski, Philip. "Dark Matter." *The Georgia Review* 54, no. 3 (2000): 431–441.
Serres, Michel. *The Birth of Physics*. Trans. Jack Hawkes. Manchester, England: Clinamen Press, 2000.

ON THE SOCIAL ORIGIN OF TIME IN LANGUAGE

Walter Schweidler

Abstract: The relation of a human person towards the whole of his or her life is paradigmatic for the nature of temporal beings in general. In ourselves as living beings we find the only possible answer to the question what is time and what it means to exist in time. We could not think of "being" as completely independent of the existence of living beings. And yet we know as persons that something has existed before any living being and will exist after the existence of any living being in the universe. This is a paradox in the self-relation of temporal beings which can be rooted in the institutional structures of human societies by which our self-experience is transformed into the symbolic representation of reality in general.

Keywords: Origin, Daseinsrelativität, Zeitgestalt, Phenomenology, Structuralism, Dasein.

Let me begin with a quotation from Heidegger which I think to be one of the deepest characterizations of the tricky relation between time and our human existence, i.e. of the relation between time and what Heidegger in *Being and Time* calls "*Dasein*"

> Of course only as long as *Dasein is* (that is, only as long as understanding of Being is ontically possible), "is there" Being. When *Dasein* does not exist, "independence" "is" not either, nor "is" the "in-itself". In such a case this sort of thing can neither be understood nor not understood. In such a case even entities within-the-world can neither be discovered nor lie hidden. *In such a case* it cannot be said that entities are, nor can it be said that they are not. But *now*, as long as there is an understanding of Being and therefore an understanding of presence-at-hand, it can indeed be said that *in this case* entities will still continue to be. (§ 43 b, 254)

The Paradox of Temporality

What is the point of this passage? Firstly, it is an expression of our everyday conviction that things exist "in-themselves," i.e. independently of any consciousness of that existence, that things will exist and events will go on after we have ceased to exist and that things have existed and events were going on before any living or conscious being came into existence.

Secondly, it directs our attention to the fact that the expression of that conviction is bound to a form that contains and connects a propositional and a non-propositional part, and that the non-propositional part presupposes what is neglected in the propositional part: our existence as living and thinking beings. That second aspect, of course, is the decisive one for Heidegger. There is what we could call a paradoxical structure in our natural view of our temporal relation to the world in which we live. In order to express our certainty that the world has existed before us and will go on without us we must, consciously or not, refer to a condition of the meaning of the word "exist" which is given only as long as we are there. So, there is a kind of self-contradiction in our consciousness of temporal existence, and the nature of that contradiction is not a logical but a performative one: when we speak about what was going on before us and will be going on without us we cannot avoid ascribing to that which was before and will be after us its relation to what is going on only as long as we exist. And the concept which keeps and covers the contradictive aspect of this relation is the concept of time. The existence of things and events beyond our own existence was and will be their existence in time; and the existence of what is going on only as long as we exist is our existence in time. So, in some sense, our time is in fact inclusive of courses of great swathes of time from which we are, normally speaking, excluded in the passage of time in our universe. We can and we do neglect that paradox in our ordinary ways of speaking, but it is present nevertheless in the hidden indexicality to our temporal experience, inherent in our speaking of what was before or will be ahead of us. Any phrase like "There was no life on earth five million years ago" or "There was no life on earth when it came into being" presupposes a frame of reference from which it makes sense for the person who utters or understands such a phrase to measure and to compare temporal relations. Wittgenstein's ironic question "How late is it now on the sun?" may be seen as a brief cross-check of this genuine interdependency between lifetime and "world-time," which is embedded in our linguistic behavior (see Blumenberg).

The philosophical problem which we face with that paradoxical structure in our understanding of temporal existence is the question of *the origin of time* itself. The aspect under which our temporal existence is constitutive for the frame of reference to any temporal being in general cannot be included in the natural flow of time that contains the beginning and end of living and conscious beings in the world. So, the difference between lifetime and world-time time tends to lead us to the assumption of an extratemporal origin of time. This tendency must not be misunderstood as "idealistic." In fact, the Kantian solution to the problem of the origin of time in the *Critique of Pure Reason* is given under the title of a "refutation of idealism" (326; B 274). Idealism, according to Kant, is defined by the claim that the existence of things outside of my consciousness cannot be proven and indeed is impossible.[1] Against this his own theorem is directed, that the "mere consciousness of my own existence ["*Dasein*"]—a consciousness which, however, is empirical in character—proves the existence of objects in the space outside of me" (B 274).[2] I become aware of myself as a multiplicity of representations given to my inner sense, and I can identify the unity of these representations in time—which is what I call myself—only in relation to something which exists permanently outside of myself. Identity in time presupposes an interdependence of change and permanence which cannot be situated totally within myself. However, just in order to defend this refutation of idealism, Kant is forced to his genuine metaphysical conclusion, i.e. the reduplication of the "I" as an empirical subject in time and the "transcendental subject" which is free of any temporal conditions. It is the dualistic essence of this dichotomy, the *chorismós* between temporal and extratemporal sphere, which, according to Kant, makes it impossible to give any rational reconstruction of the beginning or the end of the world, including the world-time. Time is not an empirical concept derived from experience but an *a priori* form of intuition; it is a "formal condition of appearances." That means that if there is any answer to the question of the origin of time, we will have to find it in the irreducible basis of our experience, the constellation between the extratemporal subjectivity and the world of objects, including the individual human being whom I call myself.

Heidegger's elucidation of the problem is representative of the whole phenomenological rejection of the Kantian "transcendental subject" and the claim to reinterpret it in anthropological terms. The Kantian idea of the "constitution" of objects by the conscious activity of an extratemporal pure subject makes it impossible to understand the origin of the conscious being itself as part of the temporal world. The Kantian strategy in dealing with that problem was the alleged "modesty" of drawing the line of demarcation between meaningful and nonsensical questions by which he claimed to have banished the question of the origin of time

1 Explicitly Kant here refers to Berkeley's position which he calls "dogmatic"—in contrast to Descartes' "problematic"—idealism.

2 Ibid. Translation here following Heidegger (who has "*Dasein*" instead of "existence").

from the scope of permitted philosophical deliberation. But through the back-door of his dualism of the temporal and extratemporal subject the whole problem reenters the center of his system. This becomes obvious in his theory of the "schematizing" of the pure intellectual concepts by temporal representations.[3] It is time what makes possible the "constitution" of objects through the activity of that intellectual subject which was designated as absolutely timeless. Where does that pre-natural time come from if not from a temporal element within that transcendental power which, by definition, cannot be thought of as included in nature. So, the paradox which we found explicated in Heidegger's remark, the paradox of our human time that includes in itself the borders of the whole process that excludes us from by far the most parts of what is going on in our universe, is back on the stage if we accept the Kantian dualism. But what is the alternative that the phenomenological approach has to offer?

The Phenomenological Reformulation of the Paradox of Temporality

The phenomenological alternative to the Kantian topic of "constitution" can best be characterized by Max Scheler's term "*Daseinsrelativität*" (relativity of being). Time is not constituted by subjective consciousness but "*daseinsrelativ*" to it, that is to say, ontologically dependent on its existence. That means that time would not be "time" if there were no conscious beings, but it does not mean that there was no content or substance of time before the origin of such beings or that time would not go on after they cease to exist. The origin of time cannot be thought of as a totally creative starting point *ex nihilo*, it must be seen as a genuine performance of transformation, an act of the turning of a pre-temporal reality into what we then consciously experience as the flux of time that encompasses the preconscious, conscious and post-conscious stages of the development of our universe. In the context of this brief presentation I can only return to the example of the experience of phenomena that contain comparative

---

[3] "Now it is quite clear that there must be some third thing, which must stand in homogeneity with the category on the one hand and the appearance on the other, and makes possible the application of the former to the latter. This mediating representation must be pure (without anything empirical) and yet *intellectual* on the one hand and *sensible* on the other. Such a representation is the *transcendental schema*" (*Critique of Pure Reason*, Transcendental Analytic, B 177, 272).

mental and material aspects, e.g., hunger.[4] When I am hungry, I become aware of the fact that I have hunger; my consciousness of my hunger is the consciousness of something that has already been there, but I cannot say that it would have been there without my consciousness of it. Given the fact that there is the present experience of my hunger, I cannot use the word "hunger" correctly without referring by it to something that is essentially conscious but that neither has been "constituted" by my consciousness nor can be reduced to it. This example is more than a metaphor: it stands in a kind of metonymic relation to the whole of my experience of myself as a *person*. If I would really become aware of myself as pure subjectivity, as the experience of a present moment in which I, for whatever reason, come to identify a chain or cluster of mental states as the one which is "mine," I could not avoid the dualism that leads to the absolute separation of the subjective sphere from the world of temporal objects. But as a matter of fact, I find myself not as a complex of mental experiences but as a person. That means: I have a body with a certain, i.e. human, nature. Becoming aware of myself means to identify myself as one human being among others and also as a living being who becomes conscious of what it has been when it becomes conscious of it. And it is this personal self-experience by which my personal existence is itself a kind of metonymic example for the whole temporal reality. That is the point of the Aristotelian concept of the "essence" of anything as *to ti en einai*: as for a person to "be a being" is to be someone who has "has been" (Spaemann 107f), so, analogically, for everything to be something and to have been what it is, is inseparable. That is, in a nutshell, the view of the origin of time which avoids the Kantian dualism but saves the aspect of "*Daseinsrelativität*": when we use the word "to be," when we speak of what things and events "are," then we are referring to a temporal relation between all of them including ourselves which would have not existed without us and yet, *given* our existence, must have existed before our beginning and will exist beyond our end. And so, the origin of time—including the time in which living and conscious beings have come into existence—is indeed going on as long as personal life is exercising what is constitutive for a person, i.e. the forming of one's life into a temporal unity. Subjectivity is what all persons share and what therefore, as Kant has seen, cannot be identified with any concrete individual being that has a beginning and

[4] I repeat the example from my speech on "Ontological Aspects of Absolute Past" in Asilomar, 2007.

end in time; but what persons do not share and what makes each one of them the temporally identifiable subject that he or she really is, is neither extratemporal subjectivity nor causally determined objectivity, but it is the form of life which in each one of us originates as long as we follow the way to it.

The Sociocultural Aspect of Temporality

But who is designated by the word "we" in this context? Who is the being whose form of life opens the horizon of the origin of time, and why is that so? Heidegger's answer has, as far as I can see, remained the core of the phenomenological dealing with this problem. His answer was: what makes it possible and at the same time necessary that my existence becomes the paradigm of being in general, is my death—or precisely: my conscious relation to it. The connection between my consciousness of what I am and the whole of my existence is the consciousness of what I will never catch up with in my life, i.e. my death. For us, personal beings, through the consciousness of our death the word "life" acquires a specific meaning which it would not have without us but which, given our existence, is also inherent whenever we ascribe the quality of a "living" being to anything. It is the meaning of "life" as a temporal form (*Zeitgestalt*). Through their consciousness of the finitude of their lives, personal beings develop a relationship towards themselves, something which in the final part of this essay I shall refer to as the "hermeneutical." As a person, one is confronted with different possible versions of the story of one's life which can be told and between which one, to a certain extent, must choose. And the code through which these versions are to be interpreted and the choices that must be made between them are constituted by one's own culture. So, for us as persons, "life" is not only a biological but at the same time a cultural phenomenon; it is the designation of our individual biography, a *vita* that includes the reflection of its *curriculum* and will always be directed by such reflection—be it our own or the reflection of others who may guide or even dictate the ways of our lives. From this point of view, from the perspective of our own personal existence, we can ascribe the character of a temporal form to every living being, independently of the question if it has or could ever have any kind of consciousness of that kind of existence. Thus it is a culturally constituted relation to ourselves which makes it possible and necessary for us to characterize natural beings—

including ourselves and our ancestors and descendents—as temporal forms. And thereby we have arrived at the principle of what I would characterize as the phenomenological alternative to the Kantian view of the extratemporal subject—which, when we keep in mind that for Kant the Newtonian concept of space and time was the indubitable starting point of any philosophical reconstruction of scientific knowledge (see Stegmüller), in its deepest core is the view of the intellect of the omniscient God who is the owner of the absolute perspective that defines the temporal structure of the universe. The phenomenological alternative to that view is the assumption that when we extend our own experience of temporality to the whole nature which we are part of and finally to the whole universe as the last horizon of nature in general, then we apply and somehow translate our experience of our personal existence as the subject of a temporal form to the world that encompasses us. Time is, so I would say again, more than a mere metaphor, the code by which we make legible and numerable, readable and calculable the universe. In time we find written the metonymic relation, the structure of the *pars pro toto*: that relation that persists between the whole of our universe and the aspect under which we are a part of it, represents it as a whole. In Leibnizian words, to be a mirror of the universe is not the privilege of the person; it is rather what we, the reasonable, share with all other beings; but what gives our specific and unique relation to the whole that encompasses us is that we have become able to read it, to listen to what it has to tell us. That is what makes time the principle of translation—in linguistic and in mathematical terms.

This is, as far as I can understand and try to explain it, the key to the phenomenological approach to the relation between time and human existence that we find indicated in the quoted passage from Heidegger. What are the criteria for its further explication and for its possible defense or refutation? In the last part of this paper I want to make in principle two remarks about this. The first is directed to the conditions of such defense or possible refutation of the phenomenological approach, the second to what I would call—certainly not with the consent of many phenomenological philosophers[5]—its metaphysical consequences.

---

[5] Concerning the relation between phenomenology and metaphysics, may I refer to my article "Das Menschenunmögliche."

The Temporal Character of Truth

Concerning the first point, I want to emphasize the methodological principle of the phenomenological approach to the origin of time, i.e. its *hermeneutical* character. The relation between personal life and the temporal structure of the universe has not been and could not have been invented by the speculation of any philosopher. What we find expressed in Heidegger's quotation and in the connecting thoughts which I tried to summarize here is the hermeneutical explication of an idea which is incorporated in the fundamental structures of human cultures and societies. It is their work, and our philosophical reconstruction is the interpretation of that work. All philosophical principles that we could ever formulate in order to show the foundation of that work depend on it ontologically and could never replace it. This is the necessary limitation of any philosophical approach to the knowledge about the nature of human life which we find embedded in sociocultural structures. On the other side, the consequence of this hermeneutical understanding of human knowledge is that philosophy is equally limited concerning the task that results from the insight into such socioculturally constituted knowledge: we can philosophically explicate and formulate that task but we cannot fulfill it. It is the task that human cultures and societies have posed to themselves and thereby to us, not as philosophers but as persons. It is the task to contribute to that fundamental sociocultural work which I have tried to characterize as the turning of a pre-temporal reality into the consciousness of time. The phenomenological alternative to the Kantian approach to the origin of time must interpret that act of turning not as a transcendental activity but as a social practice which is based on structures that are the subject of cultural anthropology. I think that the essence of these structures has been pointed out by Claude Lévi-Strauss:[6] they are structures of translation, i.e. structures of the sociocultural work by which human beings transform the structural elements of the order of nature which surrounds them into the code of communication about the tasks which lie ahead of

[6] I have always found the following rather short and unspectacular passage as the key to "structuralism": "The mistake of Mannhardt and the Naturalist School was to think that natural phenomena are what myths seek to explain, when they are rather the medium through which myths try to explain facts which are themselves not of a natural but a logical order" (95). The concept of "medium" here used by Lévi-Strauss marks exactly the alternative to the Kantian "schema" as the connecting link between untemporal ("logical") forms of thinking and our temporal reality.

them in order to shape the forms of their lives. The transformation of the names of things and events structured by the order of natural individuals and species into the names in which we find embedded the culturally ordered horizon of what we have to spend the time of our lives for must be understood as a process going on in the time which is itself constituted within this process; that is the essence of what Edmund Leach has defined as "social time"(ch. 7), i.e. the time which emerges from the process of constantly but discontinuously "rebearing" our lives in the forms of ritualized and institutionalized representations of its renewed origin. It is the time in which society organizes the cultural work of exploiting the totality of natural structures in order to offer its members the chance to enter into the perspective of the "view from nowhere" (see Nagel), the time in which we become capable to 'catch up with ourselves," i.e. to identify ourselves with the process of translation of simultaneity into succession that is going on through us, while we perceive it as the time of our universe. So, the paradox that the time from which we see ourselves encompassed as living beings in the world has its origin within our lives can at least be traced back to the sociocultural origins of "social time," i.e. the system of interruptions of our lifetime which initiates its symbolic transformation into the order of world-time.

The metaphysical consequences to which I finally want to refer are directly connected to this view of the origin of time in social practice. From our modern perspective, it seems strange that many ancient societies do not only claim but are even, in their self-understanding, based on the claim of being in a collective way responsible for the protection and the maintenance of the world (Assmann 169f). But the cultic consciousness sees collective existence as dependent on the constant renewal of the beginning "world" from which the present condition of the society and all its members is derived. The reasons for that, as far as one can try to understand them from the perspective of cultural anthropology, are not intrinsically different from our own world views. These reasons are based on the deep human consciousness of the fact that the capability which more than any other makes us human—that is to say, our thinking—is dependent on an essence of our "world" which can be understood and expressed through language. But if language is the ontological—that is, in the sense in which I spoke of the relativity of existence ("*Daseinsrelativität*") as being the ontological alternative to transcendentalism—presupposition of the *differentia specifica* of our human existence which distinguishes us from all other beings, it then follows that language cannot be understood as a product of our natural condition. Hence, language cannot have had a purely human, social, or world-immanent origin. This, at least, I would call the essence of the cultic stage of the mind, which is still indirectly present in our consciousness of the responsibility for the preservation of the world's temporal existence. There must be an original content of language that cannot be translated but only protected and saved by means of sociocultural procedures—procedures which at the same time are dependent on our cultural identity and social stability.

Again, Lévi-Strauss has pointed out that what I here call the cultic stage of our human mind is relevant for us not as a historical relic or a trace of "primitive" superstition but rather as the expression of a universal human experience. It is an experience which we can even understand as the sociocultural version of that individual relation between the content and the object of consciousness for which I above took as an example the experience of hunger. Using our language, we become aware of it as something that has always already been there when it comes to our consciousness. "Language, an unreflecting totalization," argues Lévi-Strauss in reference to Pascal's famous claim of the *raisons du coeur*, "is human reason which has its reasons and of which man knows nothing," and any system that a human subject can ever find in it rests on the presupposition that "this subject is one who *speaks*: for the same light which reveals the nature of language to him also reveals to him that it was so when he did not know it, for he already made himself understood, and that it will remain so tomorrow without his being aware of it, since his discourse never was and never will be the result of a conscious totalization of linguistic laws" (252). If this is a kind of sociocultural parallel to the experience of my hunger and to its metonymic relevance for the whole self-understanding of human persons, then we must expect here also some parallel to our certainty that hunger, although not being constituted by our consciousness of it, yet would not be hunger without this consciousness. And that is exactly the conviction which I mentioned as representative for the cultic consciousness, i.e. the conviction that as human beings we are responsible for keeping alive by our rituals and by our sacred discourse the message by which its creative principle remains embedded in the universe. Once being there, language confronts us with its connection to a ground which remains beyond and yet encompasses our horizon as its own condition of existence. That is the universal dimension of cultic consciousness confronting us "with an unconscious teleology, which although historical, completely eludes human history" (Levi-Strauss, 252).

Therefore, I think it is not a matter of romanticism when we reflect on the connection between the message of cultic rituals in which human societies claim and attempt to solve the paradox of collective identity, i.e. to repeat the origin from which they are derived and to repeat it *as origin*, as if it were present now and would be initiated for the first time, and our own self-understanding as civilized persons. When we try to give a scientific explanation of the world, to show the laws that are valid now as they were from the beginning, and to turn these laws into the again and again renewed attempt to shape the ways of our lives, then we are not absolutely remote from the fundamental condition of cultic consciousness. Repetition of the unrepeatable (the impossible) is the memory of a past that never was present catching up with oneself. We are faced with precisely the problem that, in order to understand what is going on now, we have to go back, in some sense, to the origin of the universe. This problem is the problem of time, the paradox of the relation between our present and the past that is present in it, without having been past before it. And it is the problem which I tried here to characterize on the basis of Heidegger's analysis of existence and its connection to our view of the world. I would like to end with a final quotation that marks that problem on which we will have to work when we try to understand the origin of time in our scientific language:

> *Dasein* ... is essentially in the truth ... *"There is" truth only in so far as Dasein is and so long as Dasein is* Entities are uncovered only *when Dasein is*; and only as long as *Dasein is*, are they disclosed. Newton's laws, the principle of contradiction, any truth whatever—these are true only as long as *Dasein is*. Before there was any *Dasein*, there was no truth; nor will there be any after *Dasein* is no more. For in such a case truth as disclosedness, uncovering, and uncoveredness, *cannot* be. Before Newton's laws were discovered, they were not "true"; it does not follow that they were false, or even that they would become false if ontically no discoveredness were any longer possible [...] To say that before Newton his laws were neither true nor false, cannot signify that before him there were no such entities as have been uncovered and pointed out by those laws. Through Newton the laws became true; and with them, entities became accessible in themselves to *Dasein*. Once entities have been uncovered, they show themselves precisely as entities which beforehand already were. Such uncovering is the kind of Being which belongs to "truth." (Heidegger 269)

References Cited

Assmann, Jan. "Officium memoriae: Ritus als Medium des Denkens." *Religion und kulturelles Gedächtnis*. Ed. Jan Assmann. München: C.H. Beck, 2008. 167–184.
Blumenberg, Hans. *Lebenszeit und Weltzeit*. Frankfurt am Main: Suhrkamp, 1986.
Heidegger, Martin. *Being and Time*. New York: Harper & Row, 1962.
Kant, Immanuel. *Critique of Pure Reason*. Translated and edited by Paul Guyer and Allen W. Wood. Cambridge, New York, Melbourne: Cambridge UP, 1998.
Leach, Edmund. *Kultur und Kommunikation: Zur Logik symbolischer Zusammenhänge*. Frankfurt am Main: Suhrkamp, 1978.
Lévi-Strauss, Claude. *The Savage Mind*. University of Chicago Press, 1962.
Nagel, Thomas. *The View From Nowhere*. Oxford University Press, 1986.
Schweidler, Walter. "Das Menschenunmögliche: Zur Abgrenzung von Phänomenologie und Metaphysik im Ausgang von Heidegger." *L'impossible*. Ed. Stefano Semplici. Pisa and Rome: Archivio di filosofia, 2010. 315–325.
Spaemann, Robert. *Persons: The Difference between 'Someone' and 'Something.'* Oxford University Press, 2006.
Stegmüller, Wolfgang. *Kants Theorie der Erfahrung*, in *Aufsätze zu Kant und Wittgenstein*, Darmstadt, 1970.

THE ORIGINS OF LANGUAGE AND NARRATIVE TEMPORALITIES

Rosemary Huisman

Abstract: The paper compares J.T. Fraser's modelling of nature from twentieth-century physics with M.A.K. Halliday's modelling of language in the theory of systemic functional linguistics. It demonstrates that traces of the natural human environment in which language originated can be identified in the meaning choices of language for talking about experience. Consequently, the different temporalities, or understandings of time, associated with each natural level in Fraser's model, can be associated with the different worlds of experience described in Halliday's model of language. Narrative studies have assumed that narrative is the human means of organizing our understanding of time, so it follows that the one narrative text can tell stories of different worlds with their different temporalities and, by extension, even those of modern science beyond the worlds in which language evolved. Each type of story makes temporal sense according to its own mode of coherence; the paper illustrates historical changes in the prominence of one world or another with examples from Old English to the so-called postmodern novel.

Keywords: Time, narrative, experiential meaning, English literary fiction, coherence.

Time is of the essence in relation to narrative. In response to his own question: "what does narrative do for us?" H. Porter Abbott answers, "first... many things," but adds, "if we had to choose one answer above all others, the likeliest is that: *narrative is the principal way in which our species organizes its understanding of time*" (Abbott 3. Italics in original.).[1] But it has been made explicit in the work of J.T. Fraser, founder of the International Society for the Study of Time, that the singular noun "time" is misleading. More appropriately, we could speak in the plural of temporalities, several understandings of temporal meaning, so that Abbott's comment becomes "narrative is the principal way in which our species organises its understanding of temporalities." My aim in this paper is to suggest the compatibility of Fraser's model, derived from physics, with the model of language developed by the linguist, M.A.K. Halliday. This compatibility will then allow me to turn to the study of narrative, and suggest some implications for its modelling and historical development.

[1] Similarly, Marie-Laure Ryan in "Towards a definition of narrative," in *The Cambridge Companion to Narrative* (23).

Fraser describes five nested integrative levels of nature, which I will call simply levels 1 to 5.[2] They are nested in that each emerges from the previous, with increasing orders of complexity, and in that each earlier level continues to co-exist with those which have later emerged. The levels so identified are sufficiently stable and differentiated to have characteristic properties of temporality and causation. Table 1. describes the attributes of these different levels, the upwards arrow signifying the direction of physical change and organic evolution.

Table 1. J.T. Fraser's model of the evolution of temporalities.

	Nested integrative levels of nature	Hierarchy of temporalities	Canonical forms of causation	World
5.	human "minding"	sociotemporal	collective intentionality/ historical causation	social world/ society
		nootemporal	individual long-term intentionality	mental world of individual human
4.	living matter (*organic being*)	biotemporal	short-term intentionality	physical world of living organism
3.	matter (*material being*)	eotemporal	deterministic lawfulness	inorganic physical world
2.	particles + mass (*stochastic being*)	prototemporal	probabilistic lawfulness	wave-particle world
1.	photons no mass (*becoming*)	atemporal	none—chaos	electro-magnetic radiation

Fraser uses the word *umwelt* for the spatio-temporal "reality" experienced by the inhabitants of each level. (The term was originally introduced by the biologist Jakob von Uexküll for an organism's understanding of its environment.)[3]

[2] My account of Fraser's work is drawn principally from his three books given in References. For a list of Fraser's time-related works to 2004 see *KronoScope: Journal for the Study of Time*, 4 (2004): 185–196.

[3] In *Time, Conflict, and Human Values*, Fraser defines *umwelt* as "the circumscribed portion of the environment which is meaningful and effective for a given species" (23). For

Level 1 is the world of electro-magnetic radiation, a world inhabited by photons with no mass. Their umwelt is characterised by atemporality, where everything happens at once, a world of becoming rather than being, lacking a form of causation. This is the reality of a chaotic world.

Level 2 is the world inhabited by wave-particles with mass. Fraser calls its time prototemporality, for it is the world in which the possibility of time emerges. Its causation or lawfulness is probabilistic; it is not possible to point to the exact instant when a particle is emitted, only to the probability of its having been emitted as, for example, in the half-life of radioactive material (so the half-life of cobalt 60 is 5.27 years, that is, after 5.27 years half of the possible particle emissions will have taken place). In this world, the reality of being is stochastic, possible, but not identifiable.

Level 3 is the world of inorganic physical matter, the world of galaxies, the world of material being. Fraser calls its time eotemporality, for the dawn of time. (Eos was the Greek goddess of dawn.) Its value—*t* in a physicist's formula—can be measured, though the value will depend on the measurer's spatial location. As a consequence, this temporality is not directional, is even reversible. Its causation is that of deterministic lawfulness; its inhabitants exist in a reality of determinable cause and effect.

Level 4 is the world of living organisms, of organic being, including, of course, the physiological world of the human species. Its time, biotemporality, makes sense to us in our human experience. Biotemporality is directional, moving from birth to death for the individual organism. It is the first temporality to which the concept of "the present" is relevant, as the time at which the organism feels or satisfies its need. Its canonical form of causation Fraser calls short-term intentionality, as in the organism's intention to satisfy its needs of survival and reproduction, and its umwelt (its understanding of reality) is adequate to the extent it enables that satisfaction.

The final integrative level of nature, level 5 in this nested sequence, is that peculiar to the evolution of the human species, what Fraser has referred to as human "minding." This level is associated with the most complex development, that of the human brain, a "boundary of the universe," in Fraser's poetic phrase. It gives rise to not one but two worlds of human experience: that internal to the individual consciousness, with its particular memories, predictions and fantasies, and that of social

a discussion of the term's origin in biology and its current use in Peircean semiotics, see Deely (59–62). David Herman also suggests a relevance of *umwelt* to narrative study in "Re-minding Modernism," in *The Emergence of Mind* (265–266).

interaction and consensus in myth, ritual and historical understanding. The time of the individual mental world Fraser calls nootemporality (from *noos*, Greek for mind), with the associated causation that of individual long-term intentionality. The time of the social world, the society of the human group, Fraser calls sociotemporality, with the associated causation that of collective intentionality, also described as historical causation.

From the point of view of human existence, all these five levels always co-exist, as each level is nested within the previous. Any reflection on human affairs and natural occurrences reveals that chaotic and unpredictable events still occur. However, explicit awareness of the earlier levels is recent. Fraser's model of nature is formulated from the advances of twentieth-century physics;[4] the earliest three levels, as implied in the theories of special relativity, quantum theory and general relativity, extended the human umwelt, the human understanding of reality (*Time and Time Again* 39–44). Until the end of the nineteenth century the last two levels, with their three associated worlds of biotemporality, nootemporality and sociotemporality, constituted the human umwelt, reality as understood from the experience of human life on earth.[5]

Fraser's model enables us to describe more accurately differences in the understanding of time running through two threads of narrative study, which I identify loosely as "scientific" and "philosophical."[6] The scientific thread effectively begins with the French-speaking Swiss Ferdinand de Saussure, passes via the Russian formalists and Prague structuralists to the French-speaking Bulgarian Tzvetan Todorov, and thence through French and French-influenced structuralism to its apotheosis in Gérard Genette's study of Proust. In this tradition, in a manner comparable to Saussure's *la parole* and *la langue*, the narrative is assumed to have a surface telling, in which the chronology of events may be disordered, and an underlying sense or story in which chronology is restored—as in the French dualism

[4] As a non-physicist, I find Kennedy, *Space, Time, and Einstein* particularly helpful.

[5] It is entirely outside the scope of this paper to consider the dualisms of mediaeval theology (human time and divine eternity) or Platonic Idealism (timeless truth and particular appearance).

[6] Narrative study is a garment of many threads. Ansgar Nünning maps many approaches in *Narratology or Narratologies?* James Phelan adds a chapter, "Narrative Theory, 1966–2006: A Narrative," to the revised edition of *The Nature of Narrative*. David Herman, Monika Fludernik and Brian McHale describe their versions (including denials) of histories of narrative theory in the "Prologue" of *A Companion to Narrative Theory*. I give a brief overview in *Narrative and Media* (28–44).

of *le discours* and *l'histoire*, or the Russian *sjužet* and *fabula*.[7] Thus the organising temporality of this "scientific" thread is biotemporal, the time of world 4 of living organisms. In this world, one event follows another in the organic "now" in a directed movement from birth to death.

In contrast, in the tradition of the philosophical thread, time is more typically understood as nootemporality, the time associated with the individual human mind of world 5.[8] The inter-related work of the American William James (on "consciousness") and the Frenchman Henri Bergson (on "*durée*") is significant,[9] together with Bergson's influence on the Russian Mikhail Bakhtin (Rudova). (Bakhtin himself moves into a more complex understanding of temporality in his theory of chronotopes, the different space-time characteristics of different narratives. He explicitly acknowledges his borrowing of the term "space-time" from Einstein's Theory of Relativity [84].) The French metaphysical tradition perhaps reaches its fullest expression in the work of Paul Ricoeur, in his three volume opus, *Narrative and Time*. Most recently, in this "philosophical" tradition, we see a reinterpretation of some of these authors in the context of postmodern scholarship: the American Gary Morson using particularly the writing of Bakhtin, the French Gilles Deleuze rehabilitating the study of Bergson.[10]

Now, narrative can be realised in different semiotic media, but here I am particularly interested in narratives told in language. Human language developed, I assume, in the context of Fraser's integrative levels 4 and 5: the interpenetrating three worlds of the human body, human mind and human society. Is this contextual evolution of language traceable in language? And from a study of language, what further insights might we gain into the narrative of these different worlds?

To explore these questions I turned to the particular linguistic theory known as systemic functional linguistics, initially developed by the British linguist Michael Halliday.[11] This is because its model of language, like Fraser's model of natural levels, enables talk about worlds of experience.

---

7 Gerald Prince describes the differences and conflations of the various narratological dichotomies under the entry story, in *Dictionary of Narratology* (93).

8 For a subtle account of the "literary aspects of philosophical writing," beginning with "Augustine and the 'problem' of time," see Lloyd.

9 The personal and intellectual relations of James and Bergson are described in Chapter 1 of Gunn, published in 1920, now fully accessible online.

10 For studies of Deleuze/Bergson see References under Elizabeth Grosz. Major works by other scholars here mentioned are listed by author under References.

11 From 2002 on, Halliday's scholarly articles have been progressively republished by Continuum (London & New York) as the ten-volume *Collected Works of M.A.K. Halliday*, edited by Jonathan J. Webster.

Table 2. Systemic Functional Linguistics: Stratification.

	Strata		Direction of realisation	Direction of construal
extra-linguistic	context of situation	↕	↓	↑
linguistic	semantics			
linguistic	form: lexicogrammar			
linguistic	expression: phonology/graphology			

Unlike a formal linguistics (such as Transformational Grammar), a functional linguistics does not distinguish between semantics and pragmatics.[12] In a functional linguistics, all semantics is contextually understood, that is, meaning is meaning in context, language in use, the word in the world, if you will.

Thus the strata, or levels, described in a functional description are both linguistic—the usual linguistic levels of expression, form and semantics—and non-linguistic—the level of social context. Table 2 illustrates these linguistic strata and their relationships.[13]

The direction of realisation is that of speaking or writing, of language production, world to word. The direction of construal is that of hearing or reading, of language interpretation, from word to world.

In his study of semantics, Halliday found there was a "clumping" or inter-relation of meaning choices into three groups, or "metafunctions." Halliday theorised that these three metafunctional meanings had evolved from three principal purposes humans had in using language: ideational meaning for the purpose of talking about happenings; interpersonal meaning for the purpose of interacting and expressing attitudes both to the happenings and to each other; and textual meaning for the purpose of speaking coherently in the particular situation of language use. From these three metafunctions of language, Halliday theorised three aspects of the extra-linguistic context of situation, Field, Tenor and Mode. Table 3. summarizes these context/semantics relations.

---

12 Halliday compares different types of linguistics in "Language as code and language as behaviour: a systemic-functional interpretation of the nature and ontogenesis of dialogue" (*The Semiotics of Culture and Language: Volume 1 Language as Social Semiotic* 3–35).

13 Halliday and Matthiessen point out that a more appropriate graphic metaphor is a nesting of enclosing circles: context enclosing semantics, semantics enclosing the formal level of lexicogrammar, lexicogrammar enclosing the expression level of phonology and graphology (*An Introduction to Functional Grammar* 24–25).

Table 3. Systemic Functional Linguistics: Context & Semantic Metafunctions.

Context of situation:	FIELD of social action	TENOR of situation	MODE of message
Semantic Metafunctions:	IDEATIONAL: Experiential & Logical	INTERPERSONAL	TEXTUAL
Evolution from human purposes:	to talk about happenings	to interact & express attitudes	to speak coherently

Tenor and Mode are obviously not irrelevant to wider narrative study, but it is Field which is particularly relevant to a comparison with Fraser's modelling of time. Most generally, Field is Field of social action; semiotically these Fields constitute the meaningful happenings in a culture. For a particular Field, language may be peripheral to its realisation (for example, a game of football), or language may be part of its realisation, together with other behaviours (for example, a game of bridge, with the verbal bids and actions of card play), or language may be the essential mode of realisation, as in telling a story, giving a lecture, having a conversation. This last kind of Field introduces a new complexity: a second order Field, which is the subject-matter of the story, the lecture or the conversation (Halliday, *Language as social semiotic* 143–144). It is this last doubled Field which will be relevant to narrative studies, with the first-order Field of the narrator (the social action of telling) enabling the second-order Field of the happenings of the narrative. (This doubling of teller and told is of course reiterable, as in the structure of Chaucer's *Canterbury Tales*, where a teller tells a story in which a teller tells a story in which....)

The semantic metafunction which realises Field, that of ideational meaning, has an important subdivision into experiential and logical meaning.[14] However, for the concerns of this paper it is specifically the relation of experiential meaning and Field that is most pertinent, and in particular those experiential meanings pertaining to the "system of Transitivity," represented in the following Figure 1.

[14] For example, a single clause realises experiential meaning. But clauses can be linked together (usually with a conjunction, such as *and* or *although*) to form a clause complex; the link realises logical meaning. The ideational meaning of a clause complex has then both experiential and logical components. Using the methods described in this paper, I study the use of logical meaning to tell a postmodern narrative in "Telling Time: the Temporalities of Thomas Pynchon's Postmodern Narrative."

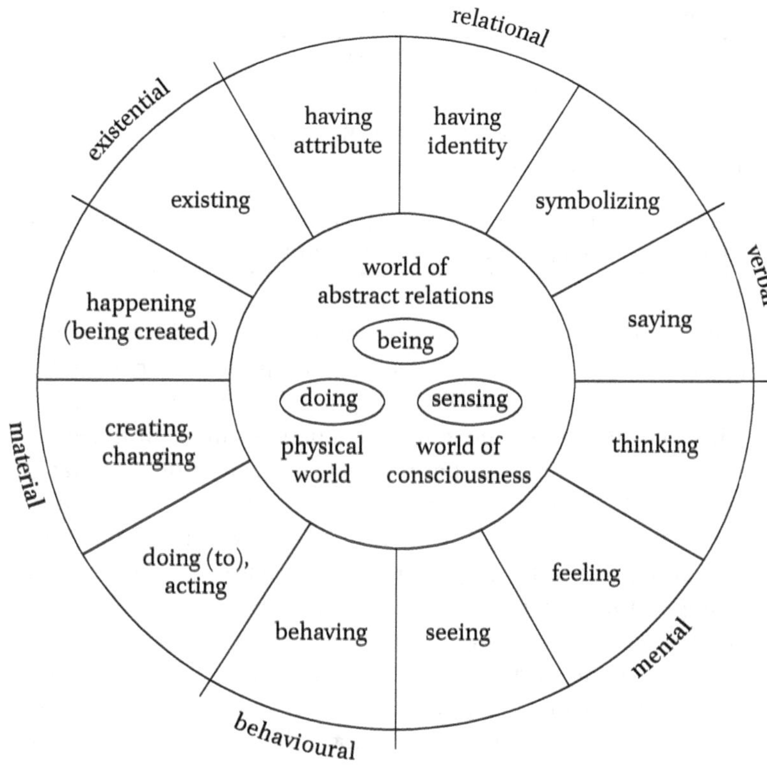

Fig. 1. The System of Transitivity.

Figure 1. appears in the body of the second and third editions of Halliday's *Introduction to Functional Grammar*, labelled "The grammar of experience: types of process in English." Demonstrating its central significance to Halliday's modelling of language, it also appeared in full colour on the cover of the second edition. The figure is a graphic metaphor for the choices of process meaning, a system of choices Halliday calls the System of Transitivity. The word "system" in systemic functional linguistics refers to choices of meaning associated with a particular unit of grammar, such as a clause, group or word. The clause is a particularly important unit for such choices, and the nucleus of experiential meaning in the clause is the semantic process, the meaning which is realised in the grammatical verb. So, to quote from the third edition,

> [the figure above] represents process types as a semiotic space, with different regions representing different types. The regions have core areas and these represent prototypical members of the process types; but the regions

> are continuous, shading into one another, and these border areas represent the fact that the process types are fuzzy categories. (172)[15]

Note particularly the reference to "core areas" and "prototypical members of the process types." The systemic terms for the process types are written on the outside of the circle. In the colour image of this figure, the three prototypical types were represented by the primary colours of red, yellow and blue: red Material processes, yellow Relational processes, blue Mental processes. Similarly, the border areas of semantic fuzziness, as one meaning shades into another, were graphically realised by secondary colours (orange Existential processes, green Verbal processes, purple Behavioural processes).

In turn, the experiential meanings of these process types are written *within* the outer circle: for the prototypical process types, Material Processes are those of *doing/acting, creating/changing, happening*; Relational processes are those of *having attribute, having identity, symbolizing*; Mental processes are those of *thinking, feeling* and *seeing*. In the following example of transitivity analysis, the lexical verb of each clause is bolded, and the process meaning it realises is printed in capitals.

From Descartes' "Meditation II":

> The Meditation of yesterday has **filled** my mind with so many doubts, MATERIAL
> that it **is** no longer in my power RELATIONAL
> to **forget** them. MENTAL
> Nor do I **see**, meanwhile, any principle MENTAL (= *understand*)
> on which they can be **resolved**; MENTAL
> and, just as if I had **fallen** all of a sudden into very deep water, MATERIAL
> I am so greatly **disconcerted** MENTAL
> as to be made unable either to **plant** my feet firmly on the bottom MATERIAL
> or **sustain** myself MATERIAL
> by **swimming** on the surface. MATERIAL (85)[16]

15 Possible experiential meanings in the clause are those of process, participants and circumstances. The choice of process determines the semantic roles (such as Actor or Senser) possible for participants, which are realised in nominal structures. Optionally, circumstantial meanings of time, place, purpose etc. may be realised in adverbial or prepositional structures. The grammar of the Transitivity System is fully discussed in Chapter 5: "Clause as representation" (168–305).

16 The linguistic phenomenon Halliday calls "grammatical metaphor" enables a process to be realised as a noun, as in the mental process "Meditation" (in line one), here used in its clause with a material process as verb (*filled*). Traditional rhetoric recognises the figurative effect of such use.

The one text will usually include choices from different process types, but one process choice may be more frequent, with an accumulation of one type of meaning. In the following example, the relational processes of identification and attribution are noticeably frequent, realised for the most part in the lexical use of the verb "to be" (bolded).

From Malory's fifteenth-century telling of the Arthurian stories, Sir Hector's threnody over the dead body of Lancelot:

> 'A, Launcelot!' he sayd, 'thou **were** hede of al Crysten knyghtes! And now I dare say,' sayd syr Ector, 'thou sir Launcelot, there thou lyest, that thou were never **matched** of erthely knyghtes hande.* And thou **were** the curtest knyght that ever bare shelde! And thou **were** the truest frende to thy lovar that ever bestrade hors, and thou **were** the trewest lover, of a sinful man, that ever loved woman, and thou **were** the kyndest man that ever strake wyth swerde. And thou **were** the godelyest persone that ever cam emonge prees of knyghtes, and thou **was** the mekest man and the jentyllest that ever ete in halle emonge ladyes, and thou **were** the sternest knyght to thy mortal foo that ever put spere in the reeste.' (725)
>
> (* passive construction for 'erthely knyghtes hande never **matched** thee', i.e. 'was never equal to you.')

As described previously, the choice of experiential meaning in the process realises the Field (direction of text production) or is construed as the Field (direction of text interpretation). Returning again to Figure 1, The System of Transitivity, we see in the centre of the circle, in the lozenge-shaped boxes, the three types of Field associated with the three prototypical types of process: Fields of *doing* with the Material processes, Fields of *being* with the Relational processes, and Fields of *sensing* with the Mental processes.

This brings us to what is central to this paper's thesis: the three worlds written adjacent to each box. These are the three co-existing worlds of human experience that can be realised by or construed through natural language: the physical world, the world of abstract relations, the world of consciousness. We understand the different construal of these domains of experience, so that we can recognise the figurative use of one domain to explain another. For example, in his "Meditation," previously quoted, Descartes uses a simile beginning "as if" to explain his Mental process in the world of consciousness (*disconcert*) by choices of Material processes of the physical world (*fall, plant, sustain, swim*).

Talk about language is talk about semantic acts, but a wider semiotic perspective, including other modes of realisation—choices of image, gesture, family structure, and so on—might lead to the following generalisation: for a given culture and language, the sum of all possible Fields of *doing* constitutes the physical world; the sum of all possible Fields of *being* constitutes the world of abstract relations; the sum of all possible Fields of *sensing* constitutes the world of consciousness. How comparable are these three semiotic worlds to the worlds of Fraser's five integrative levels of nature and, in particular, to the human *umwelt*?

For an individual human being, each theory has described three worlds of experience. The first is one external to the individual, on which s/he can act or which acts upon her/him: Halliday's physical world and Fraser's level 4 physical world of the living organism.

The second world is internal to the individual, not directly accessible to others: Halliday's world of consciousness and Fraser's level 5 mental world of the individual human. The third world is both external and internal to the individual, the world shared with the human group which yet gives attributes and identity to the individual within that shared world: Halliday's world of abstract relations and Fraser's social world. Abstract relations are exactly those relations which are socially constructed: a glance back at the extract from Malory on Sir Lancelot—analysed for the relational processes of being in the world of abstract relations—shows a catalogue of conflicting social codes, those of chivalry, romance, Christianity and more. Virtue (being *trewe*) has a different meaning in each of these different social constructions, but in the identity of Lancelot these conflicting social relations co-exist (and thereby hangs a narrative of conflicting loyalties).

I began this discussion of language with the question: is the contextual evolution of language traceable in language? where by context is meant the co-existence in the human umwelt of Fraser's integrative levels 4 and 5 in the three worlds, physical, mental and social, of human experience. In the conceptual terms of systemic functional theory, the answer is yes, for Halliday has construed these three worlds of the human umwelt from the prototypical process types in the linguistic system. In short, both theories appear to model the same worlds of human experience; the origins of language in the natural human umwelt on this planet, the human experience of physical, mental and social worlds, can be traced in the very organisation of meaning in language.

Juxtaposing the temporality (and causality) of Fraser's worlds with the meaning choices of the comparable worlds in Halliday's model enables us to tease out more fully the experiential texture of a narrative. In modern narrative theory, following the work of Genette, the word *diegesis* typically refers to the "internal world" of the narrative, the world inhabited by the characters,[17] but the discussion so far makes clear that a narrative can tell of more than one world; certainly in natural language, as it evolved, it can tell at least of the physical world, the world of consciousness, the social world of abstract relations. So just as it is more accurate to speak of narrative temporalities rather than narrative time, it is also more accurate to speak in the plural of narrative diegeses, even within the one text.

In the one narrative text, the experiential texture is produced by the weaving together of stories of the different worlds, in one way rather than another. This texture also facilitates the comparison of different narratives. My work, focused particularly on the comparison of English literary narratives, has led me to the following observation: that one can observe a different emphasis or dominance, a general difference in experiential texture, in texts from different historical periods, in different social contexts. Further, I see a generally inverse relation between the direction of natural change and organic evolution in Fraser's levels (1 to 5), and the direction of literary production in English narrative texts. The remainder of this paper explores these assertions.

Until the end of the nineteenth century, Fraser's integrative levels 4 and 5, with their three associated worlds of biotemporality, nootemporality and sociotemporality, constituted the human umwelt, reality as understood from the experience of human life on earth. Natural language had evolved to construe this experience. It is unsurprising then that pre-twentieth-century narratives weave together stories of these three worlds, though the deployment of these worlds, the interwoven texture of the narrative, varies considerably. To understand this variation, it is helpful first to consider Fraser's elaboration of "time felt" and "time understood."

Fraser makes an important distinction between the organic world, his level 4, and the mental and social worlds grouped together in level 5, the most recently evolved integrative level of human "minding." Fraser describes the temporalities of level 5 as "time understood"; in contrast he describes the human understanding of time at level 4, that of organic life, as "time felt":

[17] I discuss different uses of *diegesis* in "Narrative Concepts" (18–19). See *diegesis* in Hermann et al. (*Routledge Encyclopedia of Narrative Theory* 107–108). A wider discussion than this paper permits could develop the concept of "storyworld(s)" (see *storyworld* in Hermann, *Routledge Encylopedia* 569–570).

> The tasks of the deeper regions of the brain are the semi-autonomous controls of biological processes. They are unavailable for cognitive examination. I cannot explore the timing of my heart rate or the programming of my digestion except indirectly, through the use of instruments and theorems. This unavailability makes it difficult or even impossible to give a complete description of experiences we call feelings. (Time and Time Again, 262)

In comparison, of the human understanding of time at the level of "minding," he writes:

> The tasks of the newer regions of the brain are those of speech and associative functions. By time understood I mean the temporal dimensions of the world as perceived and hence, defined, by these newer regions. It includes our awareness of our aging and eventual death, as well as an appreciation of the temporal organization of societies and cultures. (262)

Fraser's comments on time felt and time understood are relevant to the sequential textual development in social, mental and biological stories, that is, to what follows what in the story of one world or another. The telling of each type of story has its own characteristic type of meaningful sequence, its "mode of coherence," appropriate to the temporality and causation of its world, and (from Halliday's modelling) also appropriate to the dominant prototypical meanings construing its world.

In a story of human social life ("the temporal organization of societies and cultures"), where meanings of abstract relations dominate, the narrative sequence could be called equative, because it tells of what is communally regarded as like or, conversely, what is unlike, in shared practice and value. The sociotemporality of this world is concerned not with one event after another but with the comparability of experience. From the habitual construal of this world, its persistence, a people can understand their history, myths, social rituals and genres, from the formally significant (a marriage) to the everyday (buying food).

In a story of human individual life, where meanings of sensing dominate in an individual world of consciousness, the narrative sequence is associative because it tells the successive sensations of an individual consciousness. Unlike the successive "now" of the individual's biological time felt, these sensations, subject to "cognitive evaluation," can move through an extended past and a projected future. As Fraser puts it, it includes "our awareness of our aging and eventual death" (262).

In a story of organic life, where meanings of doing dominate in an external physical world, one observable event follows another, in a narrative trajectory from birth to inevitable death. The sequence is chronological, the quantitative time of the clock in which different beings can co-ordinate their activities. Perhaps because of its public nature, as opposed to the private nature of the world of consciousness, "time" as a monovalent concept has often been equated with the biotemporality of the physical world (as, I've already suggested, in the assumptions of the "scientific" thread in narrative studies).

For each world, a story has its own characteristic type of meaningful sequence, and the terms equative, associative and chronological have been used above for the narrative sequences characteristic of social, mental and organic worlds respectively. The sequence type of a narrative world I have called its mode of coherence: it is the assumption that enables a reader/ listener to make sense of the story in terms of its construed world. A mis-assumption—possibly because of historical or cultural distance—diffuses the narrative sense. For example, as I next illustrate, scholars who assume that all stories make their sense chronologically will find the narrative of *Beowulf* lacking, even digressive. The remainder of this paper fleshes out these various assertions with examples from English narrative texts of different periods:

The narrative of the Old English poem known as *Beowulf*, written down probably in the 900s, falls into two halves. A conventional plot summary might tell you that the young Beowulf travels to the Danish kingdom to kill the monstrous Grendel and his mother, and that the old Beowulf, now king of his own people, kills a dragon but is himself killed. However the text frustrated early critics with its so-called "digressions" to stories of other kings, peoples, heroic events; an early editor famously criticised the poem for its "lack of narrative advance" (Klaeber lvii–lviii). The chronology of biotemporality, one physical event after another, is the most obvious kind of advance if you think the narrative is primarily a story of killing/physical acts (*kill* is a Material process, construing the physical world). In my reading however, published in full elsewhere, I suggest that it is the equative sequence of sociotemporality that is dominant in the text, construing the social world of the narrative in a story of social roles. To quote myself:

> Sociotemporality... is associated with a world of symbolic relations and social attributes and identities; its sequence is that by which a social group understands its history, its social being. The meaning of this narrative sequence is equative: a sequence of social similarity or, frequently, dissimilarity; it is comparable in the clause to the meaning choices of relational processes, of being or not being. ("Narrative sociotemporality and complementary gender roles in Anglo Saxon society" 126)

For example, "In... *Beowulf* the hall Heorot is raised; the narrative immediately tells us of familial strife which destroys it," and again, the Danish king "tells Beowulf he will be a comfort to his people and immediately...describes the malevolent rule of a king who brought injury to his people" (126). The chronological sequence of the physical world is occasionally dominant—notably in the underwater struggle of Beowulf and Grendel's mother, as first narrated—but much of the action is told by characters in speeches, rather than directly described by a narrator (see Huisman, "The Three Tellings of Beowulf's Fight with Grendel's Mother"). (In Halliday's Transitivity figure, Verbal processes occupy the "fuzzy" area between the prototypical Relational and Mental processes; that is, dialogue occupies an interworld between the social world and the world of consciousness.) The principal character is exemplary of a social role, rather than an individual in the modern sense. The Mental processes of the text tell of appropriate or inappropriate attitudes for a given social role, not of an individual's idiosyncratic thoughts and feelings.

In their discussion of "Representing Minds in Old and Middle English Narrative," Leslie Lockett (on Old English) and Monika Fludernik (on Middle English) provide further evidence of the less internalised understanding of mental processes in the earlier periods. Lockett maintains that "for the Anglo-Saxon audience, this mind was...literally corporeal; they thought of the mind as the part of the body responsible for thought." (44, see also 57–58). Fludernik concludes that "the radical distinction between speech and thought that has dominated narratological discussions of consciousness representation in the novel does not necessarily apply *mutatis mutandis* to medieval texts" ("1050–1500: Through a Glass Darkly" 94). David Olson gives a somewhat similar description of pre-literate Greek narrative: "The Homeric vocabulary included terms for referring to talk and feelings but whereas for their literate descendants, the Classical Greeks, such terms refer to internal, mental events, for the Homeric Greeks, they referred to external bodily, objective events. For the [latter] what we refer to as thinking is usually described as speaking, an activity originating in the organs of speaking, the lungs. Pondering is conversations addressed to those internal organs" (238). I suspect that in pre-printing narratives we are more likely to observe such traces of an oral tradition of transmission.

William Caxton is credited with introducing printing to England in the late fifteenth century, but why did he choose particular publications? For the eighteenth-century historian Edward Gibbon, Caxton's choice was obviously a response to market demands: "In the choice of his authors, that liberal and industrious artist was reduced to comply with the vicious taste of his readers; to gratify the nobles with treatises on heraldry, hawking, and the game of chess, and to amuse the popular credulity with romances of fabulous knights and legends of more fabulous saints" (qtd. in Blake 5). Initially, then, the new language technology of print reproduced the narratives of medieval inheritance, but narrative production would change as a society of print culture developed[18] and the demands of the market-place intensified.

Daniel Defoe's *Robinson Crusoe*, published in 1719, "must be one of the most popular books ever written...," reprinted and translated continuously. For editor John Richetti, its "enduring and universal appeal" is that of "an archetype of modern heroic individualism and self-reliance" (ix). Defoe himself insists on the individual authenticity of his work, carefully distinguishing *Robinson Crusoe* from the (still-popular) romances of "popular credulity."[19] In Clive Probyn's words, "As in all of Defoe's novels, the shape of this book [*Robinson Crusoe*] is determined by the shape of an individual life" (31). This developing taste for a narrative focus on the individual, superseding the dominance of sociotemporal stories, will transform the temporal texture of English literary narratives.

I earlier suggested "a generally inverse relation between the direction of natural change and organic evolution... and the direction of literary production." This is a movement from Fraser's level 5 to level 4, not a movement between the two worlds, social and mental, of level 5. The latter would imply an historical progression, from the medieval to the early modern period, of narratives dominated by the associative sequence of the mental world, but this is evidentially incorrect. The representation of consciousness certainly becomes important as the English novel develops, post-printing, but, I argue, it is a representation tied to the biotemporal progression of the individual in the chronological sequence of their lived experience. In the worlds of the natural human umwelt associated with Fraser's model of natural levels, two worlds correspond to the individual human: the mental world of Fraser's level 5, and the organic world of Fraser's level 4.

Samuel Richardson's *Pamela*, first published in 1740, has sometimes been described as "the first English novel" because of its overt concern with the individual's inner life.[20] The title page of the second edition (1741) includes the following (/ indicates lineation, // paragraph end):

[18] On the evolution of print culture generally, see Eisenstein.

[19] The title page gives Crusoe himself as the author ("Written by Himself") and Defoe's *Preface* (speaking of himself as editor) says, "The editor believes the thing to be a just history of fact; neither is there any appearance of fiction in it" (1, 3).

[20] "Richardson's undeniable contribution to the novel lay... in the literary process of creating personality through the writing of letters and the articulation of thought processes" (Probyn 61).

> Pamela:/ or,/ Virtue Rewarded./ In a Series of/ Familiar Letters/ from a/ Beautiful Young Damsel,/ To her Parents./ Now first Published/ In order to cultivate the Principles of Virtue and Religion in the Minds of the Youth of both sexes.// A Narrative which has its Foundation in Truth and Nature; and at the same time that it agreeably entertains, by a Variety of curious and affecting Incidents, is intirely [*sic*] divested of all those Images, which, in too many Pieces calculated for Amusement only, tend to inflame the Minds they should instruct.// (Keymer and Sabor 3)[21]

Richardson's explicit purpose is one of promoting the established social world ("to cultivate the Principles of Virtue and Religion"), but these principles are to be conveyed not from the heroic or holy perspective of knights and saints but from that of an ordinary, lower-class servant-girl—fictional but true-to-life ("its Foundation in Truth and Nature"). Letters, naturally in the first person, in a "familiar" context, to recipients most interested in the girl's welfare, enable the story of the events Pamela experiences, as her employer attempts to seduce her (details of the physical world) to be woven tightly with the story of Pamela's emotions, fears and convictions in relation to these events (details of the mental world). Thus (the simple communication model assumes) the mental worlds ("Minds") of individual readers will be instructed. But though "representations of consciousness" (construed from Mental processes) are noticeable in the epistolary novel,[22] the very sequence of the first person letters implies a dominant chronological sequence; the mode of coherence is that of the organic world (Fraser's world 4), as the narrative tells the succession of "presents" which constitute the directional flow of biotemporality.[23]

---

21 My transcription does not reproduce the graphic emphasis conveyed by choice of character type in size, case and font.

22 In *The Epistolary Novel, Representations of Consciousness*, Joe Bray argues for the sophistication of Richardson's examination of consciousness, objecting to the Modernists being credited with the "inward turn" to subjectivity (7–10). I can agree with Bray, yet still argue that the dominant mode of coherence is the chronological sequence of biotemporality, the individual lived life, not the associative sequence of nootemporality.

23 Of this linking thought to action, Richardson's phrase "to the moment" is much quoted: "The Nature of Familiar Letters, written, as it were, to the *Moment*, while the Heart is agitated by Hopes and Fears, on Events undecided, must plead an Excuse for the Bulk of a Collection of this Kind. Mere Facts and Characters might be comprised in a much smaller Compass: But, would they be equally interesting?" From his "Preface" to *Sir Charles Grandison* (4). Though each letter is written from memory (the world of consciousness), Richardson apparently wants his character to relive, rather than merely record, each experience and its associated emotions. Occasionally, Pamela does tell of both her mental state in the past of experience and her mental state in the present of writing: for example, in the embodied emotion of "trembled" (she writes of a failed attempt to escape, and the temptation she then felt to commit suicide): "God forgive me! but a sad thought came just then into my head!—I tremble to think of it!" (149).

The narrative of *Pamela* illustrates how, even in a novel so concerned with its character's consciousness, the external story of life lived encloses the internal story of life thought. The novels of other eighteenth-century authors, such as Henry Fielding's *Tom Jones*, are more readily accommodated to this thesis.[24] The focus is on the character as individual, negotiating his or her progress through the vagaries of the new social contexts. Though a nootemporal identity affirms the individuality of the principal character through the narrative (occasionally) telling the thoughts of that character, these thoughts are typically linked to external action. Similarly, stories of the external social world—descriptions of dress, manner, etiquette and so on, as well as place—are occasionally vivid, but their telling is subsumed to the dominant biotemporal story, as the principal character moves from one social location to another.

In the nineteenth century, we finally reach the so-called classic realist novel. I suggest this is what many people still think of as "the novel": in the classic realist text we read the most complex and coherent interweaving of stories from all three worlds, physical, social and mental, which natural language evolved to construe. George Eliot's *Middlemarch* (first published 1871–1872) is a canonical example. Eliot's narrative moves between different sequences of telling: the chronological sequence of the biotemporal physical world, the associative sequence of the nootemporal mental world, the equative sequence of the social world. The classic realist novel seems "most real" not because it describes "one world" most realistically but because it gives equal attention to describing the overlapping three worlds of everyday human experience, even though one world may appear to dominate at a given textual moment. The complexity of such texture is well illustrated in the following extract from *Middlemarch*. (I quote and paraphrase from my own previous discussion of this passage.)[25]

[Dorothea, wife of the scholarly Mr. Casaubon, is visited in her home by Mrs. Cadwallader, wife of the Rector (who had been asked to conduct a funeral),

[24] Fielding famously mocked the self-conscious virtue of Richardson's heroine in his novel of a hypocritically devious female, *Shamela* (Reproduced in Keymer and Sabor 49–118). The history of reader reception from the eighteenth century to the present of Richardson and Fielding's writing—those who favour one author over the other, at different times and for different reasons—casts interesting light on the social dimension of critical judgements (see Michie).

[25] An earlier version of my discussion of this extract appeared in "Living in different worlds" (60–63).

Sir James and Lady Chettam (who is Celia, Dorothea's sister), so they can watch the funeral from an upper window; the latter is in a room Casaubon occupied when he was 'forbidden to work' after illness, but now he is back in the library, where Dorothea too would have been except…]

But for her visitors Dorothea too might have been shut up in the library, and would not have witnessed this scene of old Featherstone's funeral, which, aloof as it seemed to be from the tenor of her life, always afterwards came back to her at the touch of certain sensitive points in memory, just as the vision of St. Peter's at Rome was interwoven with despondency. Scenes which make vital changes in our neighbours' lot are but the background of our own, yet, like a particular aspect of the fields and trees, they become associated for us with the epochs of our own history, and make a part of that unity which lies in the selection of our keenest consciousness. (Eliot, 326)

The physically external world of chronological sequence (Fraser's level 4) governs the larger structure of the novel. Thus the chapter from which this passage comes is sequentially chronological to the events of the previous chapter, in which the troubled death of the old man Peter Featherstone is narrated. At a less panoramic level within a chapter, the psychological or social worlds [of Fraser's level 5] may structure the narrative sequence.

The first sentence of the extract is within the worlds of the characters of the novel, and is dominated by the psychological world of Dorothea's consciousness, and its associative relations. In the physical world of the novel, that of chronological sequence, we are at one point of chronological time: that at which Dorothea witnesses the funeral of the old man. But from this single event the narrator moves into Dorothea's chronological future, to view the event as one associated, in future memory, with "certain sensitive points." Then, the narrator compares this future association of present event and emotion to an event already in Dorothea's chronological past, her visit to St. Peter's in Rome, and its association in her memory with an emotional state: "despondency." This is an omniscient narrator's sequence of nootemporality, telling us of Dorothea's internal world of consciousness and its habitual characteristics.

However, the second sentence of the passage encloses the biological and mental worlds of the characters within a wider social world. The narrator of *Middlemarch* assumes an explicit relationship with the reader, speaking of "we" and "us" inclusively. So now the narrator claims not only to know the psychological worlds of her characters, but to see those characters as representative of all her readers, even perhaps all humans. Paradoxically, such universality writes in a particular view of the social world—that these individual histories are written within the experience

Table 4. Six types of fictional reality.

Fraser's level	Narrative world/diegesis	Story of…	Mode of coherence	Temporality
5.	6. social	human social life	equative sequence	sociotemporality
	-------------------	-------------------	-------------------	-------------------
	5. mental	human individual life	associative sequence	nootemporality
4.	4. organic	life	chronological sequence	biotemporality
3.	3. material	being	reversible sequence	eotemporality
2.	2. uncertain	possibility	indeterminate sequence	prototemporality
1.	1. chaotic	becoming	incoherent sequence	atemporality

of a more general history. This equative conflating of individual traits to human social attributes is characteristic of a story of sociotemporality.

The last two decades of the nineteenth century and the first decades of the twentieth saw an accelerated development of the extended human umwelt—that enlarged understanding of the physical environment that Fraser describes in integrative levels 1, 2 and 3. If each world is associated with its own temporality, and temporality is the essential concern of narrative, yet natural language evolved from human experience on this earth—how could the stories of these "unnatural" extra-terrestrial worlds be told? The experimental solutions to this conundrum constitute those types of narrative which have been grouped under the terms "modernism" and "postmodernism." They augment the possibilities of fictional reality, giving us now six types of narrative world associated with Fraser's five levels, each with its own type of story.[26] It is still possible to assert that each story makes sense, according to its world's temporality, through its characteristic sequence or mode of coherence. Table 4. describes these augmented narrative realities.

What does this mean for worlds of the extended umwelt (the paradox of making coherent sense of an incoherent sequence)?

[26] It may well be that sciences other than physics can add narrative worlds to this model. For example, genetics suggests a temporality in the organic world different from that of the single organism, as Jay Clayton describes in "Genome Time."

William James' phrase "stream of consciousness" is often used to describe the associative sequence, the telling of a character's mental sensations, through which the so-called Modernist authors, such as Virginia Woolf and James Joyce, realised the internal world of an individual character, the mental life of that individual with its own nootemporality.[27] However, the narrative texture can be more complex, touching the external world through an inflecting sequence of internal and external perspectives between different characters and narrator. Thus the Modernist narrative can tell a story inclusive of eotemporality, the temporality of Fraser's level 3, as the thought processes of different characters, in their differing nootemporalities, intersect or diverge, like the movements of large matter in the space-time of Fraser's level 3, where events can be determined in space/time but can be differently measured from different locations.

In the following extract from Virginia Woolf's *To the Lighthouse*, we are variously told the thoughts and the speech of Mrs. Ramsay, the inner state of Mr. Tansley (the young man "feeling himself out of things") and Mrs. Ramsay's consciousness of that state, Tansley's (or the narrator's?) thoughts of Mrs. Ramsay ("giving out a sense of being ready"), the narrator's description of Mr. Carmichael's physical appearance in the less dominant story of the physical world... This is a world of shifting epistemology, where there is not one unambiguous site of knowledge, of interpretation, at which the reader can confidently position her/himself:[28]

> Insoluble questions they were, it seemed to her, standing there, holding James by the hand. He had followed her into the drawing-room, that young man they laughed at; he was standing by the table, fidgeting with something, awkwardly, feeling himself out of things, as she knew without looking round. They had all gone—the children; Minta Doyle and Paul Rayley; Augustus Carmichael; her husband—they had all gone. So she turned with a sigh and said, "Would it bore you to come with me, Mr. Tansley?"
>
> She had a dull errand in the town; she had a letter or two to write; she would be ten minutes perhaps; she would put on her hat. And, with her basket and her parasol, there she was again, ten minutes later, giving out a sense of being ready, of being equipped for a jaunt, which, however, she must interrupt for a moment, as they passed the tennis lawn, to ask

[27] In James' words, "Consciousness, then, does not appear to itself chopped up in bits. Such words as 'chain' or 'train' do not describe it fitly as it presents itself in the first instance. It is nothing jointed; it flows. A 'river' or a 'stream' are the metaphors by which it is most naturally described. In talking of it hereafter, let us call it the stream of thought, of consciousness, or of subjective life...." (Ch XI).

[28] Brian McHale initially associated modernist fiction with the disruption of epistemology, postmodern with the disruption of ontology in *Postmodernist fiction*.

> Mr. Carmichael, who was basking with his yellow cat's eyes ajar, so that like a cat's they seemed to reflect the branches moving or the clouds, passing, but to give no inkling of any inner thoughts or emotion whatsoever, if he wanted anything. (20–21)

Much has been written on the emergence of early twentieth-century modernist literature and its environment of influence; I am not suggesting a simple linear relation between science and literary innovation. Indeed Henri Bergson, whose ideas are seen as significant in the period, famously disagreed with Einstein, and displacements of space and time are more immediately seen in art, which in turn more directly influenced literary production.[29] In contrast, however, most postmodernist literature appears self-consciously aware of the extended umwelt of contemporary science.

Just what is literary postmodernism? There is no necessary agreement on its distinction from modernism,[30] but I would argue that narratives which include stories of Fraser's worlds 1 and 2 can be so called—the chaotic world 1 of atemporality, where everything happens at once, the uncertain world 2 of prototemporality, where you can't point to the instant when something happens, but only to a statistical possibility. The atemporal story, with an incoherent mode of coherence, undermines the very possibility of identity—of all the entities traditionally understood to constitute narrative: differentiated character, setting, event—while the prototemporal story undermines the certainty of identification.[31] In a postmodern narrative these counter-intuitive stories are (or may be) woven together with stories from all four other worlds. They are not necessarily dominant numerically in the words of the text, but their experiential meaning is so noticeable—by the modes of coherence of other worlds, they just don't make sense—that even a little of them can earn the literary label "postmodern."

---

[29] On Bergson's contribution, see Gillies, and also Douglass. On science and modernism more generally, see Schleifer, and also Vargish and Mook, as well as Whitworth.

[30] For example, Bran Nicol writes: "while [French author Alain] Robbe-Grillet is included in some classifications of the postmodern, [Leslie] Fiedler makes the point that the *nouveau roman* is too 'serious' and 'neo-neo-classical' to be called postmodern. I agree with this point—if ever there was an artist who required the label 'late modernist' Robbe-Grillet is the one—and this is why, as fascinating as I find Robbe-Grillet's pronouncements on the novel, his work is not included in this reader" (137).

[31] In Ursula K. Heise's words: "Postmodernist novels thereby project into the narrative present and past an experience of time which normally is only available for the future: time dividing and subdividing, bifurcating and branching off continuously into multiple possibilities and alternatives" (55).

Much of the writing of the American author Thomas Pynchon has been called "postmodern," including his 2006 *magnum opus* of 1085 pages, *Against the Day*. In the very subject matter of the novel, Pynchon appears to have embraced the universal, all- things-coexisting temporality of Fraser's world 1. A misleading plot summary might be the high school history topic: "the causes (& after effects) of the first world war," and the historical period covered is indeed from 1893 to about 1920. However the innumerable characters, geographical locations and social preoccupations defeat any attempt at informative plot summary. (Compare this postmodern excess with that of the modernist James Joyce in his novel *Ulysses*, which traces the minutiae of the one day of an individual rather than that of the many days of the multitude.)

In *Against the Day*, the character Renfrew recommends this world 1 perspective for making sense of history:

> "Best procedure when considering the Balkans," instructed Renfrew, "is not to look at components singly—one begins to run about the room screaming after a while—but all together, everything in a single timeless snapshot, the way master chess players are said to regard the board." (689)

Paradoxically, suggests Renfrew, in a chaotic situation, individual identity impedes "making sense." Renfrew is one of two professors, he in Britain and Werfner in Germany, whose "dangerous machinations" are much talked of.[32] It is not irrelevant to the narrative that Renfrew and Werfner are identical but reversed spellings (they are in fact the one person? or not?) and (prototemporally) other characters are never sure if they have seen one or the other. The concept of "bi-location" (the name of one section of the novel) is used to describe doubling in physical matter, in history, in human individuality—for example, with doubled identity in the stories of actual doppelgangers, the one character who participates in situations at different places at the same chronological time…

Pynchon's *Against the Day* is postmodernism "in your face," but the postmodernist narrative can lead the reader more quietly into the worlds of probability and chaos. Thus Alan Palmer describes the Epilogue of Ian McEwan's novel *Atonement* as unsettling the reader's previous identification of story and character (322–336). And though the novelist Paul Auster wrote uncompromisingly postmodernist narratives in the

32 For example, by the Grand Cohen, leader of a British order that explains History by reference to the Tarot pack. See the Grand Cohen's humorous construction of an alternative history of the Victorian era (230–231).

three novellas of *The New York Trilogy*, dominated as they are by chance and the erasure of identity, the chaotic causation of world 1,[33] his later novel *Invisible* could be read as modernist until… in retrospect, the reader's confidence in the ontology of person and event is taken away.[34]

In conclusion: human beings evolved in the context of all five nested integrative levels of nature described in J.T. Fraser's model. However, human awareness of experience took place only in the context of organic life, level 4. and human "minding," level 5. These levels were the context for the origination of human language.

Halliday's system of Transitivity describes the choices of meaning for realising (talking/writing about) or construing (understanding) human experience (hence experiential meaning). These meanings talk about three prototypical worlds of experience: the physical world, the world of consciousness, the world of abstract/social relations.

Juxtaposing Fraser and Halliday's models, we can conclude that the natural context for the origination of language is indeed traceable in the very choices of experiential meaning. The physical world construed by language enables talk about the natural level of organic life, while the worlds of consciousness and abstract relations enable talk about the natural level of human "minding." At the same time, this comparison raises a question: what of Fraser's levels of nature 1, 2 and 3, those of the extended human umwelt, which were not recognized in the human context in which language evolved?

The levels of nature differ significantly in their temporalities. But narrative has been understood to be that telling of meaning which organises time. So if human language can be used to talk about the earliest levels of nature, even though its experiential meanings did not evolve in the context of those earliest levels, then it will be in some form of narrative.

---

33 Brendan Martin, "Dislocation, Ambiguity, Indeterminacy: The Postmodernity of *The New York Trilogy*," in *Paul Auster's Postmodernity* (103–144). In "Time and Space," in Herman, Teresa Bridgeman cites Auster's *City of Glass* (first novella in the *Trilogy*) as an example of the "non-realist" text, in which the "traversing of spatio-temporal barriers is possible, and is indeed a feature of postmodern narratives where the reader's recognition of the trangression is part of the reading experience" (52). Her paper describes time in the biotemporal dualism of the "scientific" thread of narratology: the chronological sequence of events as story, the chronological reading of the text as discourse. However, from the perspective of Fraser's model of different temporalities, described in this paper, the postmodern narrative has not traversed a barrier but rather told stories of different temporal worlds.

34 As discussed in my forthcoming article, "Paul Auster's storytelling in *Invisible*: the pleasures of postmodernity."

Such a story will make sense, be coherent, in terms of the temporality and sequential mode of coherence of those early levels. Equally, a story of any level will be incoherent, or at least found lacking, if it is understood only in terms of the temporality and mode of coherence of another level. This applies as much to the narrative of *Against the Day* as to that of *Beowulf.* On its slipcover, *Against the Day* was subtitled *A Novel*, perhaps anticipating (accurately) its labelling by some critics as an "Anti-Novel" (googling provides ample references to pro- and anti-reviews). But arguments over classification of literary objects one can leave to others. Rather, the approach to experiential texture offered in this paper enables one to see clearly why such an argument might occur, and in what kinds of narrative world this argument or that has most investment.

References Cited

Original Works Mentioned

Auster, Paul. *Invisible.* New York: Picador, 2009.
Beowulf and the Fight at Finnsburg. Ed. Fr. Klaeber. Boston: D.C. Heath, 1922.
Defoe, Daniel, *Robinson Crusoe.* 1719. Ed. John Richetti. London: Penguin, 2001.
Eliot, George. *Middlemarch.* 1871–72. London: Penguin, 1994.
Fielding, Henry. *The history of Tom Jones, a foundling.* 1749. London: Collins, 1974.
Malory, Works. Ed. Eugene Vinaver, 2nd ed. Oxford: Oxford University Press, 1971.
Pynchon, Thomas. *Against the Day.* London: Jonathan Cape, 2006.
Richardson, Samuel. *Pamela.* 1740. London: J.M. Dent, 1914.
——. *The History of Sir Charles Grandison.* 1753–54. London: Oxford University Press, 1972.
Woolf, Virginia. *To the Lighthouse.* 1927. London Routledge, 1994.

Works of Criticism and Theory

Abbott, H. Porter. *The Cambridge Introduction to Narrative.* Cambridge: Cambridge University Press, 2002.
Bakhtin, M.M. *The Dialogic Imagination, Four Essays*, Ed. Michael Holquist, Trans. Caryl Emerson and Michael Holquist. Austin: University of Texas, 1981.
Bergson, Henri. *Time and Free Will, An Essay on the Immediate Data of Consciousness.* 1889, authorised translation by F.L. Pogson. 1910. London: George Allen and Unwin Ltd., 1950.
Blake, N.F. *William Caxton and English Literary Culture.* London: The Hambledon Press, 1991.
Bray, Joe. *The Epistolary Novel, Representations of Consciousness.* London: Routledge, 2003.
Bridgeman, Teresa. "Time and Space." *The Cambridge Companion to Narrative.* Ed. David Herman. Cambridge: Cambridge University Press, 2007. 52–65.
Clayton. Jay. "Genome Time." *Time and the Literary.* Ed. Karen Newman, Jay Clayton and Marianne Hirsch. New York and London: Routledge, 2002. 31–59.
Deely, John. *Basics of Semiotics.* Bloomington: Indiana University Press, 1990.
Descartes, René. *A Discourse on Method.* 1637. London: J.M. Dent, 1912.
Douglass, Paul. *Bergson, Eliot and American Literature.* Lexington, KY: Kentucky University Press of Kentucky, 1986.

Eisenstein, Elizabeth L. *The Printing Revolution in Early Modern Europe*. Cambridge: Cambridge University Press, 1983.

Fludernik, Monika. "1050–1500: Through a Glass Darkly; or, the Emergence of Mind in Medieval Narrative." *The Emergence of Mind, Representations of Consciousness in Narrative Discourse in English*. Ed. David Herman. Lincoln and London: University of Nebraska Press, 2011: 69–100.

——. "Histories of Narrative Theory (II): From Structuralism to the Present." "Prologue" to *A Companion to Narrative Theory*. Eds. James Phelan and Peter J. Rabinowitz. Oxford: Blackwell, 2005. 36–59.

Fraser, J.T. *Time and Time Again, Reports from a Boundary of the Universe*. Leiden and Boston: Brill, 2007.

——. "List of publications." *KronoScope: Journal for the Study of Time*, 4 (2004): 185–196.

——. *Time, Conflict, and Human Values*. Urbana & Chicago: Illinois University Press, 1999.

——. *The Genesis and Evolution of Time, A Critique of Interpretation in Physics*. Amherst: Massachusetts University Press, 1982.

Genette, Gérard. *Narrative Discourse, An Essay in Method*. 1972. Trans. Jane E. Lewin. Ithaca, NY: Cornell University Press, 1980.

Gillies, Mary Ann. *Henri Bergson and British Modernism*. Montreal & Kingston: McGill-Queen's University Press, 1996.

Grosz, Elizabeth. *Time Travels, Feminism, Nature, Power*. Sydney: Allen and Unwin, 2005.

——. *The Nick of Time, Politics, Evolution and the Untimely*. Sydney: Allen and Unwin, 2004.

——, ed. *Becomings, Explorations in Time, Memory and Futures*. Ithaca and London: Cornell University Press, 1999.

Gunn, Alexander J. *Bergson and his Philosophy*. New York: Dutton, 1920.

Halliday, M.A.K. *Collected Works of M.A.K. Halliday*. Ed. Jonathan J. Webster. London & New York: Continuum, 2002–2007.

——. revised by Christian M.I.M. Matthiessen. *An Introduction to Functional Grammar*, 3rd ed. London: Arnold, 2004.

——. *An Introduction to Functional Grammar*, 2nd ed. London: Edward Arnold, 1994.

——. "Language as code and language as behaviour: a systemic-functional interpretation of the nature and ontogenesis of dialogue." *The Semiotics of Culture and Language: Volume 1 Language as Social Semiotic*. Eds. Robin P. Fawcett, M.A.K. Halliday, Sydney M. Lamb and Adam Makkai. London: Frances Pinter, 1984. 3–35.

——. *Language as Social Semiotic. the social interpretation of language and meaning*. London: Edward Arnold, 1978.

Heise, Ursula K. *Chronoschisms: Time, Narrative, and Postmodernism*. Cambridge: Cambridge University Press, 1997.

Herman, David. "1880–1945: Re-minding Modernism." *The Emergence of Mind, Representations of Consciousness in Narrative Discourse in English*. Ed. David Herman. Lincoln and London: University of Nebraska Press, 2011. 243–272.

——. "Histories of Narrative Theory (I): A Genealogy of Early Developments." "Prologue" to *A Companion to Narrative Theory*. Eds. James Phelan and Peter J. Rabinowitz. Oxford: Blackwell, 2005. 19–35.

Herman, David, Manfred Jahn and Marie-Laure Ryan, eds. *Routledge Encyclopedia of Narrative Theory*. London & New York: Routledge, 2005.

Huisman, Rosemary. "Paul Auster's storytelling in *Invisible*: the pleasures of postmodernity." *Storytelling: Critical and Creative Approaches*. Eds. Jan Shaw, Liam Semler and Philippa Kelly. Houndmills, Basingstoke, Hampshire UK: Palgrave Macmillan, forthcoming.

——. "Telling Time: the Temporalities of Thomas Pynchon's Postmodern Narrative." *Semiotics 2009; "The Semiotics of Time."* Eds. Karen Haworth, Jason Hogue and Leonard G. Sbrocchi. New York, Ottawa, Toronto: Legas, 2010. 243–252.

—— "Narrative sociotemporality and complementary gender roles in Anglo-Saxon society: the relevance of *wifmann* and *wæpnedmann* to a plot summary of the Old English poem *Beowulf*." *Journal of the Australian Early Medieval Association* 4 (2008): 125–137.
——. "Living in different worlds—the possibilities of English literary narratives." *Forum for Musings: The JNTU Journal of English Studies* 2.2 (2008): 55–80.
——. "Narrative Concepts" and "From structuralism to post-structuralism." *Narrative and Media*. Eds. Helen Fulton with Rosemary Huisman, Julian Murphett, Anne Dunn. Cambridge: Cambridge University Press, 2005. 11–44.
——. "The Three Tellings of Beowulf's Fight with Grendel's Mother." *Leeds Studies in English* N.S. XX (1989): 217–247.
James, William. *The Principles of Psychology*. 1890–1891. New York: Cosimo, 2007. 18 July 2012. <http://webspace.ship.edu/cgboer/jamesselection.html>.
Kennedy, J.B. *Space, Time and Einstein, an Introduction*. Montreal & Kingston: McGill-Queen's University Press, 2003.
Keymer, Thomas and Peter Sabor, eds. *The Pamela Controversy, Criticisms and adaptations of Samuel Richardson's* Pamela *1740–1750*. London: Pickering & Chatto, 2001.
Lloyd, Genevieve. *Being in Time, Selves and Narrators in Philosophy and Literature*. London and New York: Routledge, 1993.
Lockett, Leslie. "700–1050: Embodiment, Metaphor and the Mind in Old English Narrative." *The Emergence of Mind, Representations of Consciousness in Narrative Discourse in English*. Ed. David Herman. Lincoln and London: University of Nebraska Press, 2011. 43–68.
Martin, Brendan. *Paul Auster's Postmodernity*. New York: Routledge, 2008.
McHale, Brian. "Ghosts and Monsters: On the (Im)Possibility of Narrating the History of Narrative Theory." "Prologue" to *A Companion to Narrative Theory*. Eds. James Phelan and Peter J. Rabinowitz. Oxford: Blackwell, 2005. 60–71.
——. *Postmodernist fiction*. New York: Methuen, 1987.
Michie, Allen. *Richardson and Fielding, The Dynamics of a Critical Rivalry*. Lewisburg: Bucknell University Press, 1999.
Morson, Gary Saul. *Narrative and Freedom: The Shadows of Time*. New Haven and London: Yale University Press, 1994.
Nicol, Bran, ed. *Postmodernism and the Contemporary Novel*. Edinburgh: Edinburgh University Press, 2001.
Nünning, Ansgar. "Narratology or Narratologies? Taking Stock of Recent Developments, Critique and Modest Proposals for Future Usages of the Term." *What is Narratology?* Eds. Tom Kindt and Hans-Harald Müller. Berlin & New York: Walter de Gruyter, 2003. 238–275.
Olson, David. *The World on Paper: the Conceptual and Cognitive Implications of Writing and Reading*. Cambridge: Cambridge University Press, 1994.
Palmer, Alan. "1945–: Ontologies of Consciousness." *The Emergence of Mind, Representations of Consciousness in Narrative Discourse in English*. Ed. David Herman. Lincoln and London: University of Nebraska Press, 2011. 273–297.
Phelan, James. "Narrative Theory, 1966–2006: A Narrative." *The Nature of Narrative, Revised*. Eds. Robert Scholes, James Phelan, Robert Kellogg. Oxford: Oxford University Press, 2006. 283–336.
——. "Narrative Judgements and the Rhetorical Theory of Narrative: Ian McEwan's *Atonement*." *A Companion to Narrative Theory*. Eds. James Phelan and Peter Rabinowitz. Oxford: Blackwell, 2005. 322–336.
Prince, Gerald. *Dictionary of Narratology*. Lincoln and London: University of Nebraska Press, 2003.
Probyn, Clive T. *English Fiction of the Eighteenth Century, 1700–1789*. London: Longman, 1987.
Richetti, John. "Introduction." *Robinson Crusoe*, Daniel Defoe. London: Penguin, 2001. ix–xxviii.
Ricoeur, Paul. *Time and Narrative* Vol. 1–3. Trans. Kathleen Blamey and David Pellauer. Chicago: The University of Chicago Press, 1984–1988.

Rudova, Larissa. "Bergsonism in Russia: The Case of Bakhtin." *Neophilologus* 80 (1996): 175–188.

Ryan, Marie-Laure. "Towards a definition of narrative." *The Cambridge Companion to Narrative*. Ed. David Herman. Cambridge: Cambridge University Press, 2007. 22–35.

Schleifer, Ronald. *Modernism and Time. The Logic of Abundance in Literature, Science, and Culture, 1880–1930*. Cambridge: Cambridge University Press, 2000.

Todorov, Tzvetan. "What is Literature For?" *New Literary History* 38 (2007): 13–32.

Vargish, Thomas and Delo E. Mook. *Inside Modernism, Relativity Theory, Cubism, Narrative*. New Haven: Yale University Press, 1999.

Whitworth, Michael. *Einstein's Wake: Relativity, Metaphor, and Modernist Literature*. Oxford: Oxford University Press, 2001.

SECTION TWO

FROM FUTURE TO ORIGIN: TEMPORAL INVERSIONS AND ASYMMETRIES

NARRATIVE FICTION: WRITING TOWARDS THE ORIGIN

Sabine Gross

Abstract: In telling stories, one can employ a range of ways to deviate from basic temporal sequence. Fictional narratives present a particularly complex set of strategies and conditions that fundamentally rearrange and scramble surface trajectories of temporal and causal elements. With reference to Edgar Allan Poe and a number of other authors and texts, the article investigates the paradoxical loops that mark fictional representations of sequence and of causality in a confounding of origin and destination. It discusses the "double logic" of realist texts that successfully camouflage the structure of reverse causality by an overlay of plausibility, and proceeds to highlight the specific role of detective fiction in establishing a self-conscious second-order model of "writing backwards."

Keywords: Narrative, temporality, realism, detective fiction, German literature, Edgar Allan Poe.

I. *Textual Trajectories*

The writing and reading of fictional narratives generates a specific relationship between the concepts of 'origin' and 'future,' one that throws some of the most significant and intriguing aspects of narrative fiction into sharper relief. Fictional texts have origins, as of necessity anything that exists must: in their conception, temporally, and intratextually. In the first instance, they originate in the author's ideas, initial impulse, and plans; their temporal origin is the moment or phase of inception; and in intratextual or intradiegetic[1] terms, they can be said to originate in a propitiating event or encounter, the initial thrust from a state of equilibrium into one of imbalance upon which all narratives depend. But what does it mean to say that they have 'futures'? Here, I beg a certain degree of license to extend the term beyond its strictly temporal meaning: formally, to the final pages of a text or episode; in terms of its content, to the intradiegetic culmination and resolution of the narrative; and finally, to its intended

[1] Referring to the fictional world established in a story or film, pertaining to the experienced reality of the characters. For this and other narratological terms, Abbott provides a fine basic introduction; Herman et al. offer extensive and authoritative coverage.

destination, that is, the actualization in the mind of the reader that brings it to life and transforms it from (printed) words to a representation of characters and events.[2] Sequentiality provides the most fundamental principle by which to draw these elements into the structure implied by origin and future. Lives progress from birth to death, from origin to ending, in linear fashion. Speech is subject to the same linearity. At first glance, the same seems unambiguously true for the direction and progression of written narratives: words are arranged sequentially—on the page, in the book or journal, or on the screen—and we process them in sequence, generally reading narrative fiction from beginning to end, first line and page to last. As we read, we follow a story in which developments and events similarly progress from origin to endpoint. Narrative texts, however, actually establish a dual temporal order, "two kinds of time and two kinds of order," as Abbott (14) puts it. The "double time structuring" (Chatman 118) of tale and telling, of events and narrative, of narrated time (*erzählte Zeit*) and time of narration (*Erzählzeit*)[3] have, in narratological terms, been captured in numerous paired distinctions: *fabula* and *sjužet*, story and discourse, *histoire* and *récit*, to name but the most prominent.[4] One of the marvelous features of human language is that it allows us to deviate from the order of events in their linguistic representation by way of using tenses and other temporal markers—adverbs and adjectives as well as designations of time, day, and year. Christa Wolf begins her memorable personal account of the 1986 Chernobyl disaster, *Accident/A Day's News* (1989) with a sentence in which the grammatical tense of the future past yokes together origin and future, what has not happened yet and the way in which it would affect a future look back at the day she records: "On a day about which I cannot write in the present tense the cherry trees will have been in blossom" (5). (*"Eines Tages, über den ich in*

---

[2] Sternberg (163–164) argues for the use of "future" in two senses, within "textual-figurative and mimetic-narrative past, present, and future." His figurative use is implicitly oriented towards the reading process: "whatever awaits us in the sequel is 'future' in relation to that point upon the continuum of the text at which we find ourselves at any given moment—the 'present' point." In the "mimetic-narrative" sense, he considers past/present/future as "literal, chronological terms" (164)—tacitly taking the vantage point of either characters or narrator based on the temporal referentiality of the narrative.

[3] These two terms go back more than sixty years, to Müller 1948. "Erzählzeit" or "time of narration" is, of course, if not a misnomer, an imprecise category for written text. As a virtual construct, it is largely equivalent to the time of reading as measured by the number of pages/length of text.

[4] Martinez and Scheffel (26) offer an authoritative synoptic overview, listing nearly twenty pairings and multi-part distinctions.

der Gegenwartsform nicht schreiben kann, werden die Kirschbäume aufgeblüht gewesen sein" [9].) In narratives, this translates into an array of more complex temporal divergences between story and discourse, although the extent to which texts can play with the presumed sequence comes up against both aesthetic and logical limits, as occasional attempts at radically reversing endpoint and beginning show.[5]

The narrator's options for complicating the simple progression from origin to endpoint[6] have equivalents on the part of the reader, beginning with the physical shape of the text and our access to it. No matter whether one reads a material object such as a book or magazine, or via an e-book on an electronic reader: once we have moved beyond the material scroll and its insistent linearity, all of these written texts allow the reader to skip and jump, in her/his turn diverging from the suggested sequence of reading. We can read the ending first; jump ahead intermittently, either to check out future developments or perhaps to select and follow one of several narrative strands; or return to earlier passages to remind ourselves of details or savor them in light of subsequent plot turns.[7] That this kind of liberty taken with the suggested sequence changes the experience of reading is obvious; it can generate delicious recognition of how the author prepares the outcome, but also a loss of two affects/effects. Once we know the outcome, we have relinquished surprise, our response to an unexpected event. Suspense—one's response to an anticipated, but uncertain

5 Cases in point: Martin Amis's *Time's Arrow*, which is told from the death of the narrator to his birth; Ilse Aichinger's "Spiegelgeschichte"; and a 2008 cinematic attempt, *The Curious Case of Benjamin Button*, directed by David Fincher, whose screenplay was based on F. Scott Fitzgerald's short story of the same title.

6 Narrators can complicate narratives by taking numerous other liberties, many of which affect their reliability. The slippery but productive concept of the "unreliable narrator," originally and famously introduced by Wayne Booth (158) as a narrator who neither "speaks for [n]or acts in accordance with the norms of the work," has since morphed into a much more complexly developed and contested category. The capsule definition in Herman et al. (623), while referring to Booth, takes the term significantly beyond its origins, and a wealth of narratological treatments (for a useful summary, see Nünning in Herman et al., 495–497) attest to its possible scope. Here, a brief mention must suffice of the degree to which a narrator's lack of reliability (as designed by the author and/or perceived by the reader) may intersect with the temporality of narratives as well as their plausibility, drawing readers' attention to gaps or gaping inconsistencies or—conversely—allowing them to attribute such occurrences to a narrator shaped by the author to be deliberately sloppy, obtuse, or devious.

7 See Gross (*Lese-Zeichen*), esp. "Text als Lesefläche." The situation is different for live performances such as oral narration and theater: here, listeners and spectators are bound by the actual temporal progression. Only in recorded form do performances release the recipient from this linearity and allow similar freedom of access and choice of sequence as do printed or electronic texts.

event—is a different matter. It has been argued that in re-reading, suspense can be maintained even when one knows the outcome of the text, a finding borne out by readers' experiences.[8] Suspense, then, is a particular aesthetic effect generated by the text. We experience it by committing ourselves to being future-bound in a way that is not literally temporal. In re-reading, it hinges on our willingness to insert ourselves into a double virtual arc extended between origin and future: the fictional diegesis, and our re-entry into its realm as if we were encountering it for the first time. Richard Gerrig has coined the term "anomalous suspense" for this "as if" and provides empirical-cognitive data for its existence (although we may quibble with a term that seems to relegate the human propensity to enter and accept the temporal arc of a narrative to the anomalous). We may know the outcome outside the diegetic world, but within the text and its "games of make-believe" the outcome is not known, the future inaccessible (79).[9] In contrast, Caroline Levine, in an analysis that is not psychological-empirical, but sociocultural, if not anthropological, takes as her point of departure John Ruskin's *Stones of Venice* and his insistence that "partial knowledge [is] the only proper human condition..." (8). Placing suspense in the context of human (in this specific case, Victorian) "knowledge seeking," she emphasizes the extent to which it is tied more generally to "imagining a future... that despite our best guesses, may or may not come to pass..." and points out how in this very suspension, we may "experience a vital, vibrant *pleasure*" (9). For Levine, in this thrilling anxiety about whether expectations or wishes will come to pass, readers experience "a humility that is itself curiously pleasurable" and amounts to "the joy of self-suspension" (9). Narrative suspense, we could say, allows us to experience the daunting uncertainty of the future in a safe, fictional setting that domesticates the effect of not knowing and allows us to replicate it as often as we wish. Its delight is situated precisely in the partial overlapping of actual and virtual-fictional futures, the constitutive tension of categorical difference and experiential transferability.

Returning to the question of sequence: there are other, less tangible ways in which the anticipated conclusion of a narrative may serve as the origin of the act of reading—that is, the decision to read a book in the

---

[8] See Köppe and also Carroll.

[9] Gerrig as well as Levine employ dramatically contrasting approaches to the phenomenon of suspense: where Gerrig (whose chapter 5 addresses "anomalous suspense" in detail) offers empirical subjects; Levine considers readers as historical, gendered selves.

first place. Think of genre conventions for a moment. What are they but advance knowledge, or at least confident expectation, of how a narrative is going to end? Indeed, the rules of genre—and here we should widen our scope to include dramas since after all they do tell stories, if without the mediation of a narrator—tend to suggest, if not mandate, specific outcomes. In a tragedy, the protagonist is bound to tumble into disaster; in comedies, a happy ending is required that may or may not take the form of one or more marriages; in romances, final happiness of this kind is essential and readers would feel duped if heroine and hero went their separate ways at the end. In detective stories (more on which later), we expect a crime and the subsequent successful identification and capture of the culprit by a detective, with a specific imbalance of knowledge characterized by Meir Sternberg as follows: "We know in advance that the criminal is to be apprehended in the last chapter and that the detective is sure to come to no harm; but we are kept in ignorance all along of the circumstances leading up to the crime, particularly the identity of the perpetrator" (179–180).

Expectations of intradiegetic future, of specific narrative outcomes and resolutions, may acquire a moral tinge beyond a generic plot dynamic. Justice must be done, scoundrels—at least in most traditional texts—will meet their deserved end; the virtuous and noble character (male or female) whose suffering we have followed with bated breath will be elevated and rewarded: otherwise readers may well feel cheated. This trajectory is essential in fairy tales, whose redemptive ending casts its soothing virtual light "back" over all the injustices and tribulations to which the protagonists have been subjected: we follow their pursuits while anticipating the reversal of fate yet to come. Fate is equally at work in tragedy, where we know the imperfect hero's tragic flaw (*hamartia*) must lead to his/her eventual fall. In order to justify the kind of closure mandated by genre conventions and/or reader expectations, the ending inevitably shapes preceding events and in particular affects the kind of character presented to us from the beginning—a character, that is, shaped in accordance with the requirement that he or she will deserve and give meaning to the eventual outcome.[10]

[10] This is why Aristotle points out in Book XIII of his *Poetics* that tragedy must refrain from presenting either "the spectacle of a virtuous man brought from prosperity to adversity" or "the downfall of the utter villain." The only satisfactory character is "a man who is not eminently good and just, yet whose misfortune is brought about not by vice or depravity, but by some error or frailty" (45).

The logic of sequence and cause-and-effect is not only at work within texts; it straddles the separation of external and intratextual world. In a temporal loop that we should not dismiss as trivial (and to which we will return later), before we start reading the author has composed and completed the text, whose ending thus precedes our beginning to read it. In fact, the relevance of this can extend beyond the boundaries of an individual volume, as two recent instalments in Laurie King's series of Mary Russell/Sherlock Holmes mysteries confirm—to pick an example from the genre that will become our main focus later. *The Game* (2004) features a small number of apparently minor circumstances and characters—a balcony collapsing, a woman on a ship inquiring about the identity of the main protagonists—that first seem to have significance, but eventually turn out to be red herrings in the resolution of the volume's narrative. But only for this volume: in the next one, *Locked Rooms* (2005), they reappear to acquire pivotal importance, leading us back to these previous developments and reactivating them for the current and future plot. Moreover, King's texts find their origin in other, earlier texts: her justly acclaimed series constitutes a second-order fictionality within the life of a character—Sherlock Holmes—who is fictional to begin with. And what are we to make of texts that create fictional interpolations in the lives of actual historical personages (mysteries featuring, for instance, Jane Austen or Gertrude Stein as protagonist-detectives)? Here we find authors writing 'towards' an original life that is neither fictional nor intratextual, calling up readers' knowledge of these actual lives. In reading, we go along with the author's invention of developments beyond the actual lives that are simultaneously inserted into a foregone historical past.

Texts offer other reversals of origin and outcome. If we widen our view again beyond prose texts proper to include stories crafted for the stage, we find that so-called analytical dramas (as opposed to the 'regular' kind of drama that is not marked by any qualifier), by definition, do not present events as they take place, but gradually unveil what has already happened, arriving finally at a destination that is simultaneously an origin. Sophocles' *Oedipus the King* is a classical and oft-cited example (as is Heinrich von Kleist's *The Broken Pitcher*, the early-nineteenth-century comedy that takes its cue from Sophocles in featuring a judge who is eventually forced publicly to uncover his own transgressions): all of the crucial plot ingredients, twists and turns have already occurred, if we take the play's opening as reference point. The play moves towards fully unfolding retroactively how and why Oedipus got to this point: married to his mother, having killed his father, he has fulfilled the very prophecy his parents had tried to avoid by casting him out to die.

Just like analytical dramas, prose narratives may also proceed analytically. Frequently, this is effected by way of a tale of discovery or revelation inserted into the narrative world and featuring an intradiegetic narrator. Here as well, the tale's conclusion may return us to the origin of the story in various ways. A metaleptic twist may move us across different levels of the story world if, for instance, the final sentence repeats the opening sentence/s, but this time framed as the speech of the narrator as he starts to offer an account of all that has happened in the story.[11]

II *Textual Logic*

Even with the qualifications stated so far, by and large it would nonetheless seem hard to challenge the sequential presentation/occurrence of both tale and telling, and especially the temporal and/or causal progression of events presented from beginning to end.[12] Sure, to establish the order of putative events presented, the reader has to unscramble these so-called anachronies (to use Gérard Genette's [1983] widely accepted term). So what: humans are pretty good at that. And yet, one of the vexing questions about fiction that animate, in particular, theoretical discussions of *mimesis*[13] is whether tales are constituted or *re*constituted from their telling, with the '*re*' marking a crucial difference in our conception of fiction and its relationship to reality. Where in everyday life (at least when speaking in the past indicative), events precede their telling, in fiction

---

[11] This is Martin Walser's surprising move at the end of his novella *A Runaway Horse/Ein fliehendes Pferd*.

[12] "[H]ow shall I begin?" asks Oskar Matzerath in the opening chapter of Günter Grass's *The Tin Drum*, and continues with a disquisition on temporal and other conventions of novel-writing that is highly appropriate for a text flagrantly in violation of many of these conventions (5). Here, Grass offers a tongue-in-cheek response to an even more famous passage on narrating in time and of time, penned by none other than Thomas Mann at the beginning of Chapter Seven of his *Magic Mountain*. For more on the challenges to "keeping the story straight" and Oskar's role as unreliable narrator, see Hall, esp. 108–111 and Gross ("Narration in *The Tin Drum*").

[13] Variously rendered as "imitation" or "representation," the term *mimesis* has played a central role, ever since Aristotle introduced it in this function, in defining what it is that art is and does. For reliable guides through the history of the term and its uses in scholarship, see, for instance, Eusterschulte (2001), the entry in the most recent reincarnation of the *Reallexikon der deutschen Literaturwissenschaft* (Erhart), and Gebauer/Wulf).

there *are* no events that would have occurred prior to and independently of the telling: they exist "entirely as a matter of the reader's inference and construction from the text or discourse" (Shen 567). It is part of the very illusion they create that fictional narratives generally give the impression of recounting events that have already taken place or (less frequently) are taking place: in other words, that appear to precede and exist independently of a telling in which they actually originate.[14]

After the preceding, somewhat cursory, remarks about origins and fiction, it is this observation that launches us in this section on a more sustained argument about the significance of a model of origins and endings/futures for the analysis of fictional texts, continuing (in III, below) with particular attention to the genre of detective fiction. Especially in tales of discovery and revelation, the semblance of progression (even if told out of sequence), coupled with our expectations towards specific and meaningful outcomes, can simultaneously reverse the causal logic of the text such that the outcome becomes a guiding force for what precedes it. Returning to *Oedipus* for a moment: if we follow the drama closely, we are bound to realize that Oedipus's murderous guilt is never proven conclusively. Instead, the eyewitness is repeatedly quoted as speaking of many robbers, whereas Oedipus was alone.[15] The play does not ultimately resolve what in our contemporary understanding is a major contradiction and crux in establishing Oedipus's guilt.

Jonathan Culler agrees that we are never given proof. He argues that this is irrelevant, and that in fact "our conviction [of Oedipus's guilt] does not come from the revelation of the deed," asserting rightly that "we know that Oedipus must be found guilty, otherwise the play will not work at all" (174). But his observation is not limited to the ways in which the force of expectation already referred to above can guide and shape our readings. He posits that these expectations extend to the characters *in* the text: "Oedipus, too, feels the force of this logic.... His conclusion is based... on... the demands of narrative coherence. The convergence of discursive forces makes it essential that he become the murderer of Laius, and he yields to this force of meaning.... Here meaning is not the effect of

[14] A deeper engagement with the question of the reality status of fictional worlds and events would take us too far afield here. Shen argues that on the basis of its "mimetic component," the story could indeed "be taken as a non-textual given" (567). See also Ronen as well as the succinct summary by Ryan.

[15] Oedipus explicitly addresses the discrepancy and argues that "if he uses the same number, it was not I who killed [Laius]. One man cannot be the same as many" (Sophocles 147).

a prior event but its cause" (174). It is this reversal that for Culler not only obviates the need for proof, but indeed reframes the murderous act on the attribution of which the play purportedly hinges: "Oedipus becomes the murderer of his father not by a violent act that is brought to light but by bowing to the demands of narrative coherence and deeming the act to have taken place" (174).

What Culler has termed the "double logic" of the text amounts to a confounding of origin and destination and the establishment of a reverse causality. Both are at work not only in highly literary, artistic works, but hold for fiction in general—and indeed for non-fictional narratives as well. Consider the narratives we tend to construct about our own lives or those of others, for specific situations or the whole trajectory of a life, for individuals or for more than one generation. There is a distinct desire to attribute events, character features, successes or failures to preceding developments or traits to which the latter can then be traced back. Here, futures retroactively shape origins. Biographies as well as autobiographies make for intriguing reading in this respect. The question of origin can be addressed *ex negativo*: "At that point, no one could have guessed that…" ("she was destined for great things"/"20 years later he would receive the Nobel Prize"). Yet frequently, authors of biographies and autobiographies alike are intent upon discerning—or fabricating—possible origins to match subsequent developments: "In the playground sandbox, her talent as a sculptor first became evident" or "Even at this tender age, he demonstrated an extraordinary knack for words." Future developments not only precede, but are instrumental in creating their own origin: it seems highly desirable that they *be* preceded and prefigured by their own traces.

But in fiction, causality jumps levels and sequence more radically. For the author, the effect—the outcome of a plot development—is frequently the *cause* of what precedes it in the text, and her/his work proceeds in reverse. Events, characters, and details are introduced with a view towards the role they play for the future development and resolution of the plot, so that the "virtual future" presented in the text turns out to be the origin of its own foundation in a paradoxical loop. So-called prolepses—jumping ahead of the natural chronological sequence—only scratch the surface of this paradox. They are an author's way of making explicit this double structure of narratives: remarks such as "At this point, no one could have guessed that two months later, he would be dead;" an ominous "Her facial expression would come to haunt her visitor in the weeks to come;" or a suspense-inducing "Had she only known what was going to happen three days later" conspicuously activate a reader's sense of the double temporal structure of a narrative. But even without such deliberate temporal mini-excursions on the part of the narrator, texts are sprinkled with seeds of further developments and final outcomes that, I would argue, simultaneously support/stabilize the natural-seeming progression and undercut/destabilize it.

In "Freud's Masterplot," Peter Brooks, discussing endings, refers to Sartre's *La Nausée* and a passage where Roquentin memorably describes how in narrative "the end is there, transforming everything." Brooks emphasizes the "anticipated structuring force of the ending" that provides temporal shape to the narrative and sums up: "The beginning in fact presupposes the end" (1035).[16] But the insight predates the twentieth century. In "The Philosophy of Composition," his famous 1846 essay on the process of writing his even more famous 18-stanza poem "The Raven," Edgar Allan Poe polemically counters the image of the artist as genius "compos[ing] by a species of fine frenzy—an ecstatic intuition" with that of a skilled craftsman who starts from the end in order to arrive at the beginning (675). In his essay, Poe states categorically that "every plot, worth the name, must be elaborated to its dénouement before anything be attempted with the pen. It is only with the dénouement constantly in view that we can give a plot its indispensable air of consequence, or causation ..." (675). Poe proceeds to depict the alleged process in which he composed his popular poem as calculated throughout and governed by "precision and rigid consequence" (677): in this, he simultaneously sketches a sequence that starts not with the first word or stanza, but with determining the desired length, "tone," and "effect"—with the latter thus becoming the cause of the poem's form rather than its result. Having settled on a tone of melancholy, Poe explains that in then casting about for a refrain "it would have been absolutely impossible to overlook the word 'Nevermore.' In fact, it was the very first which presented itself." The challenge now lay in devising a "pretext" for its use, in "inventing a sufficiently plausible reason for its continuous repetition" (679). To summarize: Poe settles on the raven, and then the topic of death and a bereaved lover, and then he "first established in [his] mind the climax, or concluding query." The first words he writes of the poem are those of the penultimate stanza: "Here then the poem may be said to have its beginning—at the end, where all

[16] Compared with the later and slightly revised incarnation of this article as chapter 4 of *Reading for the Plot* (1985), the temporal contours of this dynamic are more clearly delineated in the 1977 version.

works of art should begin..." (680). Working backward, he then decides where to place the main character, how to introduce the raven, and other features of the text's progression. We may note in addition that the very choice of the name "Lenore" for the late-lamented love of the lyrical I combines what Roman Jakobson has defined as the essence of the "poetic principle"[17] with the kind of reverse determination that runs counter to everyday experience and the temporal arrow of lived experience: where in real life a name is primary, linked to one's identity, chosen before birth or shortly thereafter, in fiction it would be chosen to resonate with the character of the protagonist as the author imagines him or her. In Poe's case, it is assigned as a consequence of—and rhyme for—the previously selected word "Nevermore" and its poetic-affective qualities.

At the outset, Poe's provocative and almost theatrical account of this reverse construction of the text introduces the notion of writing "backwards" (675) by way of referring to a note from Charles Dickens written to Poe.[18] It is the reverse of narratives of authorship that have the creator of the text following the pursuits of their protagonists, claiming they never know in advance where their characters will take them. An instance of the latter is Günter Grass's account of writing *The Tin Drum*. Grass explains in the final minutes of his Nobel Prize interview that he had intended to give the novel's protagonist (demonically innocent Oskar Matzerath, who decides to stop growing on his third birthday) a sister, but that his fictional creation would not tolerate it: "...but he protested. There was no space/place for a sister!" (... *aber der hat protestiert. Da war kein Platz für eine Schwester!*)[19]—contrary to the elaborate plans that, as aficionados of Grass's work know, the author tends to draw up for his

17 As Jakobson famously formulated it in "Linguistics and Poetics": "The poetic function projects the principle of equivalence from the axis of selection into the axis of combination" (71)—or, to put it less technically: in literary works, especially poems, words are chosen not (just) for their meaning, but also for their formal-poetic features of sound, shape, or rhythm such as alliteration, assonance, length, word stress, or other qualities that link them on the aesthetic rather than the semantic level.

18 The opening paragraph of "The Philosophy of Composition" reads in full: "Charles Dickens, in a note now lying before me, alluding to an examination I once made of the mechanism of 'Barnaby Rudge,' says—'By the way, are you aware that Godwin wrote his "Caleb Williams" backwards? He first involved his hero in a web of difficulties, forming the second volume, and then, for the first, cast about him for some mode of accounting for what had been done'"(675).

19 The sister, metamorphosed into Tulla Pokriefke, instead assumes a significant role in the second and—even more so—the third volume of Grass's so-called "Danzig Trilogy," *Cat and Mouse* (1961) and *Dog Years* (1963). All translations from Grass's Nobel Prize speech by SG.

novels delineating all plot developments in advance. With a mix of exasperation and approval, Grass refers to his "paper characters" coming alive and developing a will of their own: "In the process of writing, the paper characters come to life and voice their dissent..." (*Im Schreibprozess werden die papierenen Figuren lebendig, widersprechen...*), and he finally states: "I had to bow to the will of the fictional character.... That is the wonderful thing about the process of writing, that which animates it." (*Ich musste mich dem Willen der fiktiven Figur beugen.... Das ist das Wunderbare, auch das Lebendige, im Schreibvorgang*).[20] (*Interview*). Poe's highly fictionalized account contributed to exposing what came to be termed the "intentional" and the "affective fallacy,"[21] whereas Grass expresses somewhat facetiously an experience of novelistic creation that he shares with numerous other authors. The two authors present complementary models of literary creativity: writing "backwards" and "forwards," calculated planning and processual exploration. No doubt actual writing practice is most frequently a blend of both, a crisscrossing of the virtual temporal fabric of fictional creation.

The *pro*gression of Poe's poem is so effective precisely because its end preceded its beginning. We see that how the author gets to that end is very different from how we as readers get there.[22] Then again, most readers will insert at least a small dose of "reading backward" in any act of reading. Speaking generally, our experience of causality in everyday life comes in two shapes. In one, we act in the expectation that our action will have a certain result, confidently anticipating an effect to follow what we define at the outset as its cause. In the other, awareness or attribution of a cause-effect relationship only develops after the fact, inferring causality and retroactively designating a cause to what we—also retroactively—now realize is (or may be) an effect. We tend to attribute causality to sequence, as expressed in the well-known "post hoc, ergo propter hoc"

[20] The term Grass uses here, "Schreibvorgang," is a synonym for "Schreibprozess" (process of writing). A literal translation, however, yields something closer to "the moving forward of writing," which fits in well with the overall argument presented here.

[21] Terms originally coined in 1946 and 1949 by Wimsatt and Beardsley.

[22] Re-reading, already mentioned briefly above, is a different story: while in re-reading a text we can give ourselves over to renewed suspense (if not surprise), we can also consciously follow the way the author has laid out our reading trajectory. Given our knowledge of the end, we are in effect reading the ending into its origins and vice versa simultanously.

("after, and therefore because of").[23] Narratives shift this human tendency to infer, discern, or create causality from the pragmatic to the aesthetic realm, where it is used in the service of reading pleasure.

For the reader of fiction, causality develops sequentially, from apparent cause to apparent effect. Yet at the same time we only attribute causality in hindsight to plot elements that at first may not appear significant. Events have to be motivated and made plausible as they occur, based on what has *gone before*, and within the realist paradigm, any deviations from this semblance of plausibility are perceived as shortcomings (the author inserting information clumsily, a failure of craft). And yet with all these aspirations to a natural-seeming flow of events, we know nonetheless that we are reading a prearranged text and not experiencing life in its multifaceted chaos. There is a reason why realistic literature, which in effect imposes order and significance on this chaos of life, provides a unique kind of reading pleasure.[24] Aiming to appear 'like life,' realist narratives are at the same time distinctly unlike life: they promise shape, significance, a judicious selection and arrangement of elements, all in the service of imposing a special order whose beginning and end, unlike the mere temporal sequence of lived reality, follow more demanding and aesthetically gratifying compositional laws.[25] While as readers we cherish that appearance of significant information as apparently coincidental, we simultaneously expect confidently that in most of the things the writer brings up so casually, there is an aesthetic intelligence at work, and that we will be rewarded later as their significance will be revealed; that what happens first has already been determined by what happens later.

"Realism" can be defined as the successful camouflage of this structural determination by an overlay of verisimilitude and apparent coincidence

---

[23] Admittedly, both philosophy and rhetoric consider this a flawed way of reasoning, warning against the errors this kind of inference can produce; and medical research struggles valiantly to disentangle the possible confusion of sequence and causality. Yet in terms of the human ability and eagerness to make sense of the world, it is an effective cognitive heuristic, and one readily available for writers, who often factor it in.

[24] The following observations leave aside avant-garde and experimental literary texts that challenge or flagrantly violate those writing conventions that allow the text to appear as a "window to the world," since they generally do not aspire to developing their narrative naturally and unobtrusively. My focus here is on texts whose combination of verisimilitude, plausibility, and richness of detail places them in the realist tradition.

[25] For attention to the kinds of structural composition and symbolic arrangement that take realist texts beyond verisimilitude without destroying that impression, see, for instance, the introduction and contributions in Kontje. For the ambivalences and aporias that attend realism's representation of reality, see Arndt (2009).

as well as plausibility: it is the extent to which "writing backward" is made to appear as "writing forward" or as an apparently artless manner of establishing the sequence of the text.[26] Here, a note on the idea of "motivation" is in order, since it applies doubly to our topic. In its everyday sense, "motivation" as applied to realist texts refers to a character's expressed or underlying reasons for certain actions. In the sense developed by Viktor Shklovsky,[27] the word instead acquires a technical-compositional meaning. For Shklovsky, "motivating" an element in the text is a way of camouflaging its formal necessity by thematic integration or inserting elements of psychological likeliness: of hiding the structural scaffolding of the text under a realist fabric of simulated life. The element in question must appear unremarkable at the time as one reads through the text, then acquire significance as we come to later events that may have been precipitated by what went before.

Even if we docilely progress from beginning to resolution in a text, the temporal and the causal coexist, intersect, and generate each other. In fiction and nonfiction alike, the causal chain from origin to conclusion may be tightly constructed; then again, causality can be attenuated, implied, even largely neglected and replaced by "mere" temporal sequence. Yet in all of this, writers appear to shape sequence into causality while actually starting out with the demands for causality that the textual arrangement will then have to satisfy once all the elements are in place. Fictional narratives thus represent a complex variant of the double temporal-causal movement from beginning to resolution and resolution to beginning. This calls, not least, for a reformulation of E.M. Forster's classical definition of "plot": "The King died, and then the Queen died of grief" (93). We can specify that "in order for the Queen to die of grief, first the King will have had to die." Killing off the king without appearing to do so merely to allow the queen to die is the art of realism, which in essence blends both senses of "motivation," the psychological and the technical. Using the outcome

---

[26] Even within the realist paradigm, though, there are numerous exceptions to this appearance of artless self-construction of a narrative. Most frequently, they involve a highly reflexive narrator looking back onto events, texts with a conspicuous split between a narrating self and a previous experiencing self, or one or several framing stories with a layering of additional intradiegetic narratives. Examples in German-language literature from the classical realist period that come to mind: Wilhelm Raabe's 1891 novel *Stopfkuchen*, Theodor Storm's 1888 novella *Der Schimmelreiter*.

[27] For example in his analysis of Sterne's *Tristram* Shandy (147–170). See also Erlich (ch. 13, "Style and Composition").

as origin is what enables writers both to insert essential elements and to hide them within the diegetic fabric of the narrative.

III. *Reading in Detective Mode*

In narrative fiction, the logic of endings not only determining, but indeed serving as origins is most conspicuously at work in detective fiction of any kind. The detective genre is strongly analytical, since in most of its subgenres, uncovering what has happened before is at least as important as any actual progression of events. In other forms of fiction, events are not necessarily told in the order in which they are said to have taken place. In mysteries, they are *necessarily not told* in that order, as in analytical drama. *Oedipus* has repeatedly been defined as the first detective story in literature. For philosopher Ernst Bloch, the Oedipus myth is "archetypal" in its "criminological knot/complication" (*kriminologischer Knoten*), and he calls the Oedipus myth "*the* primal matter of detecting," (*Urstoff des Detektorischen schlechthin*) (47–46).[28] Given how prominently *Oedipus* figures in Aristotle's *Poetics*, does Aristotle offer additional resources to enlighten us about mysteries? In an early, clever, and entertaining foray into this conjunction of classical drama and the lowly genre of detective fiction, in a lecture given in 1935 Dorothy Sayers attributes to him with tongue-in-cheek hyperbole "a theory of detective fiction so shrewd, all-embracing and practical that the *Poetics* remains the finest guide to the writing of such fiction that could be put, at this day, into the hands of an aspiring author" (179). For our purposes, her remarks on the social impact of detective fiction, on catharsis, on reversals, and on length are of less import. But she links Aristotle to precisely the double aspect of motivation in the technical formalist sense that I have already outlined earlier when she quotes him as "maintain[ing] that dreadful and alarming events produced their best effect when they occurred 'unexpectedly', indeed, but also 'in consequence of one another' " (179).

We should not hastily equate the causality of tragic effect in mythic narratives (no one would ever be in the slightest doubt about the end of the Sophoclean "analysis"; no one should doubt that it hinges decisively

28 One should note that Bloch's unorthodox coinage of "das Detektorische" elevates the individual act of detecting from mere sleuthing to a broadly conceived philosophical principle of discovery. On *Oedipus* and detective fiction, see also for instance, ch. 2 of Priestman (16–35).

on the development of the protagonist) with the genre conventions of mysteries that reduce "fate" to the realm of blood spatter and ballistics and feature characters happily devoid of development. Yet mysteries add specific moves to the dance of analytical discovery, of the simultaneous and mutually complementary combination of intentional placement and ingeniously created appearance of coincidental introduction that defines realist writing. Even while reading the text for the first time—if we are willing to emulate the detective and read in "detective mode" rather than consume the text passively—our experience is one of reading backwards as we proceed through the text; reading backwards, moreover, on two levels: that of intradiegetic motivation: "oh, that's why the character was shown earlier to act this way," and that of pursuing the way the author is weaving her or his web: "oh, that's why the author put that conversation in earlier." While in mysteries we know that certain elements are significant, indeed crucial for the resolution of the puzzle that the crime constitutes, we also know—and expect—them to be artfully hidden among other elements that are cunningly endowed with apparent significance by the author, but will inevitably turn out to be red herrings. So here the significance may well lie in their *not* being significant, and the dénouement may hinge on the least-conspicuous detail. In his 1929 account of "Sherlock Holmes and the Mystery Story," Shklovsky describes this additional dimension in structural terms in his disquisition on the detective story with recourse to Chekhov's often-cited law: "Chekhov says that if a story tells us that there is a gun hanging on the wall, then subsequently that gun ought to shoot.... This principle in its usual form corresponds in reality to the general principle of art. In a mystery novel, however, the gun that hangs on the wall does not fire. Another gun shoots instead" (109–110).

To the usual realist camouflage of structure by a variety of rich details and content elements, mysteries add both the heightened urgency of the analytical plot (the necessity to solve the crime, reaffirm the possibility of justice, and restore social order) and the additional layer of garden paths and red herrings intended to lure the reader into a web of motivations intended to deceive him or her as to the eventual significance of certain aspects, disclosures, pointed or throwaway remarks, and the like. They lift into consciousness readers' familiarity with genre conventions by generating repeated mental rewrites (of the type "the butler is always the murderer—so he can't be the murderer, it's too obvious—but then again, the author will assume that the reader will rule out the butler as too obvious, so he may be the murderer after all"). Mysteries are written backward by necessity—and they playfully appeal to readers' awareness of the extent to which this is the case. Invariably, they are read backward as well as forward, in a constant forming of new hypotheses and rearranging of evidence already presented (textual and criminological) into new patterns—at least if we accept the detective in the text as the reader's fictional counterpart. Indeed, not only do mysteries turn readers into detectives, they simultaneously present detectives as readers. Peter Hühn has persuasively defined the detective story as a double process of reading and writing: the criminal writes a misleading text in attempting to escape detection; the detective is charged with deciphering an ambiguous text, based on the confidence that there *is* indeed an unambiguous meaning—that is, an identifiable culprit and crime, a separation of story and discourse. Hühn points out that "classical detective fiction is *based* on the premise that a story and its presentation in discourse *are* distinguishable and that extracting the 'true' story from the (invariably distorting) medium of discourse is a feasible as well as valuable enterprise" (452). He then takes this central insight to a second level: "The usual constellation of story and discourse... occurs twice over: the story of the crime is mediated in the discourse of the detective's investigation; and the story of the detective's investigation, in its turn, is mediated in the narrator's discourse" (452). Two kinds of competition are taking place: the first unfolds between the criminal and the detective, one as writer, the other as reader of traces to be deciphered to solve the crime. In the second, the actual reader competes with the detective. Faced with the same information, the same clues[29] as the detective, can the reader beat the detective in ascertaining what happened, establishing the coherence in what appears to be a muddled and corrupted text?

Here, the reader may have to work against the so-called "Watson" figure, the detective's sidekick or assistant who is a frequent character, named after the first such figure in Arthur Conan Doyle's Sherlock Holmes stories. This character, when he or she serves—not infrequently—as narrator, tends to convey a muddled and misleading version of events, intended to sidetrack us or devised as a challenge for our powers of decoding and detecting. The Watson character's structural role as foil and subversive writer within the text is usually motivated by well-meaning naïveté and limited intelligence—in the 1920s Shklovsky, the first writer to define the narratological contours of this figure (106), calls him the "eternal fool"

29 At least according to specific rules of fairness that are considered foundational and faithfully adhered to in certain genres of detective fiction—see also next note.

(105). As the detective's alleged assistant or apprentice, he is both our friend and our adversary in the process of uncovering the truth.

Moreover, in classical whodunits the police frequently are cast as incompetent readers. To return to Poe for a moment: three of Poe's short stories follow this model, falling into the category of "tales of ratiocination—of profound and searching analysis," as an 1845 review of his *Tales* (believed to have been written largely by him) defines them (673): "The Murders in the Rue Morgue," "The Purloined Letter," and "Mystery of Marie Roget." Their characterization as "inductive" (although "adductive" might be the better term) amounts to precisely the reversal of origin and effect we have already witnessed in Poe's account of the writing of "The Raven"—and indeed the review states casually that "The Murders in the Rue Morgue," "[l]ike all the rest [of the 'tales of ratiocination'] is written backwards" (674). The plot of "The Murders in the Rue Morgue," to focus on a single example, presents, as it were, a rather implausible dressing-up of a contrived puzzle or—phrased in terms of realism—an utterly extraordinary occurrence. It is a text written in answer to the problem of creating an impossible murder that nonetheless has a solution. A mysterious double murder is investigated that seems to transcend human capabilities. The brilliant detective figure, Dupin, puts together a number of baffling clues correctly and deduces—showing up the police and freeing an innocent suspect in the process—that the murders were perpetrated not by a human being, but by an escaped orangutan.

Poe primes his readers with Dupin's theory and demonstration of a 'method.' Almost the entire first half of the text is taken up by a conversation between the proto-detective, Dupin, and his sidekick that lays out how puzzles are to be approached. A remark by Poe in a letter to his friend Philip Pendleton Cooke, August 9, 1846, goes to the core of what I have outlined as the dynamic between writing forwards and backwards:

> You are right about the hair-splitting of my French friend [Dupin]:—that is all done for effect. These tales of ratiocination owe most of their popularity to being something in a new key. I do not mean to say that they are not ingenious—but people think they are more ingenious than they are—on account of their method and *air* of method. In the 'Murders in the Rue Morgue', for instance, where is the ingenuity of unravelling a web which you yourself (the author) have woven for the express purpose of unravelling? The reader is made to confound the ingenuity of the supposititious Dupin with that of the writer of the story. (684)

Phrased more generally, what would be amazing coincidences in real life are nothing but acts of meticulous planning and ingenious (or perhaps less ingenious, in Poe's and Doyle's somewhat formulaic tales) strategies of motivation.

In mysteries, the hints of writing backward that all acts of reading potentially include become fully explicit: active readers will engage in a sustained mental attempt to "reverse-engineer" the text, to read it both in terms of how it successively presents developments and in terms of how much of this successive presentation is actually determined *a posteriori*. With Culler, we can "argue that every narrative operates according to this double logic, presenting its plot as a sequence of events which is prior to and independent of the given perspective on these events, and at the same time, suggesting by its implicit claims to significance that these events are justified by their appropriateness to a thematic structure" (178). While for Culler this double logic is mainly one of realistic surface (character psychology, plausibility) vs. "thematic structure" (the logic of the text qua text), it does have a strong temporal dimension and extends to the origins and endpoints of texts. In classical detective fiction especially (but not only) in the British tradition, this linking of eventual outcome to previous developments is taken quite seriously. In his "Ten Commandments for the Detective Novel" Stefan Brockhoff[30] asserts a somewhat hyperbolic quantitative tyranny as Rule no. 1: "All mysterious events that have happened in the course of the novel must be explained and resolved at the end. If at the beginning 10 burglaries, 20 kidnappings, and 30 murders take place, then at the end 10 burglaries, 20 kidnappings, and 30 murders must have found their resolution" (177; my translation). Brockhoff's rules and other analogous ones are one of two representations of the classical mystery novel as an ultimately closed system. The other is William Stowe's, who terms the classical British-style detective tale "semiotic," by which he means that it is a closed, stable web of signifiers that are waiting for the detective—working backward to what happened—to decode them. Rationality reigns on all levels, quantitative and otherwise. True—the waters of detection can get muddied by the occasional new corpse or minor plot development; but the analytical character of the text is dominant, and ultimately there is little development. Stowe contrasts this type with the more complex "noir"-style detective story of the Chandler and Hammett bent that he calls "hermeneutic," since the latter involves both

30 "Zehn Gebote für den Kriminalroman" ("Stefan Brockhoff" is really the pseudonym for Oskar Seidlin, Dieter Cunz, and Richard Rene Plant, writing mystery novels collectively in 1930s Basel/Switzerland). See also, for instance, the section "The Rules of the Game" in Haycraft.

detective and reader in a story whose trajectory is affected, its horizon of interpretation changed by the involvement of the detective. This version inevitably combines movement both forward and backward in the text, or (in Hühn's terms) the text of the crime/s keeps getting rewritten as the detective gets entangled in it.

There is a specific sense in which the mystery genre is writing not just towards the origin—which all fictional texts do to a varying degree—but writing towards a misleading origin that the reader is made to infer. Sayers points out the logic as she surveys the different kinds of "discovery" that Aristotle enumerates as possible avenues to the dénouement. Among them are several (by material signs and tokens; by memory; by reasoning) that pertain to the mystery story, but she singles out first his category of "Discovery through Bad Reasoning by the Other Party"[31] and then jumps to Aristotle's praise of Homer for his use of paralogism (185). Paralogism is the faulty attribution of a cause or origin, based on the reasoning that since A leads to B, B would necessarily presuppose A. We might add to Sayers's characterization of detective fiction, with Aristotle, "the Art of Framing Lies" (185) as sending us back to false origins.

Writers and literary theorists display a conspicuous affection for the mystery genre—Bertolt Brecht, Siegfried Kracauer, Walter Benjamin and others have read and written about mysteries; the feminist literary scholar Carolyn Heilbrun wrote a series of them under the pseudonym Amanda Cross. For literary theorists, detective fiction is a wonderful adventure playground for excursions into semiotic, hermeneutic, and epistemological territory—see, for instance, seminal contributions by Carlo Ginzburg, Umberto Eco, Thomas Sebeok, and Jean Umiker-Sebeok.[32] Hühn points out that for intellectuals, the status that the mystery novel awards readings and texts, its "valorization of interpretive order" (464), is bound to be gratifying.

While I would not dispute that position, I believe that the attractions of the genre include its elevation to prominence of what is a basic feature of all literary texts: the way in which they are constructed simultaneously forward and backward, establishing temporal and causal links between initial developments and final resolution that both follow and run coun-

---

[31] Butcher's translation (Aristotle 61) has "false inference," and "generating false inferences" (95) for "paralogism."

[32] Vogt and Most/Stowe provide outstanding collections of texts that straddle the divides among literary theory, philosophy, semiotics, psychology, anthropology, and other disciplines.

ter to the way we experience time in real life. Hühn claims that readers generally do not return to a mystery once they have read it, but concedes: "Rereading a detective novel is, however, a revealing experience with respect to the structures outlined: one can then clearly distinguish between the texts and authors and read the two stories (that of the crime and that of its detection) separately and at the same time" (459). Mysteries not only provide readers with the option either to get caught up in the spider-web of traces and clues, or to unravel it from a critical distance. They also offer the opportunity to do both at the same time: reading forward and backward.

Detective stories, then, are a genre in which the productive tension between origin and ending, cause and effect, motivation and behavior, complication and resolution, crime and solution, mystery and explanation is particularly charged. Here, the double temporal/causal logic of texts that is to some extent a feature of all literary writing plays a constitutive role. It is foregrounded and helps define the genre—a genre that manages simultaneously to confirm and unsettle, to employ and uncover the workings of this double logic in the mind of the reader. All acts of reading include—be it in unconscious acknowledgments or conscious appreciation—some awareness of the ways in which what comes later has determined and provided the originating impulse for what has come before: detective fiction is "writing back to the origins" to the second power.

References Cited

Abbott, Porter. *The Cambridge Introduction to Narrative.* Cambridge: Cambridge University Press, 2002.

Aichinger, Ilse. "Spiegelgeschichte." *Frauen im Mittelpunkt.* Ed. Patricia Herminghouse. New York: Suhrkamp, 1987. 43–51.

Amis Martin. *Time's Arrow Or The Nature of the Offense.* New York: Harmony, 1991.

Aristotle. *Poetics. Aristotle's Theory of Poetry and Fine Art.* Trans. S.H. Butcher; Introd. John Gassner. New York: Dover, 1951.

Arndt, Christiane. *Abschied von der Wirklichkeit. Probleme bei der Darstellung von Realität im deutschsprachigen literarischen Realismus.* Freiburg: Rombach, 2009.

Bloch, Ernst. "Philosophische Ansicht des Detektivromans." *Der Kriminalroman. Poetik—Theorie—Geschichte.* Ed. Jochen Vogt. München: Fink, 1998. 38–51.

Booth, Wayne. *The Rhetoric of Fiction.* Chicago: University of Chicago Press, 1961.

Brockhoff, Stefan. "Zehn Gebote für den Kriminalroman." *Wachtmeister Studers erste Fälle.* Friedrich Glauser [1936]. Zürich: Arche, 1991. 177–180.

Brooks, Peter. "Freud's Masterplot. Questions of Narrative." *Yale French Studies* 55/56 (1977). Reprinted in *The Critical Tradition. Classic Texts and Contemporary Trends.* 2nd edition. Ed. David Richter. Boston, New York: Bedford/St. Martin's, 1998. 1033–1044.

——. *Reading for the Plot. Design and Intention in Narrative.* New York: Random House, 1985.

Carroll, Noel. "The Paradox of Suspense." *Suspense, Conceptualizations, Theoretical Analyses, and Empirical Explorations.* Ed. Peter Vorderer et al. Mahwah, NJ: Lawrence Erlbaum, 1996. 71–91.

Chatman, Seymour. *Story and Discourse. Narrative Structure in Fiction and Film.* Ithaca: Cornell University Press, 1978.

——. "What Novels Can Do that Films Can't (and Vice Versa)." *On Narrative.* Ed. W.J.T. Mitchell. Chicago: University of Chicago Press, 1981. 117–136.

Culler, Jonathan. *The Pursuit of Signs. Semiotics, Literature, Deconstruction.* Ithaca: Cornell University Press, 1981.

Erhart, Walter. "Mimesis." *Reallexikon der deutschen Literaturwissenschaft.* Vol. II. Eds. Harald Fricke et al. New York: de Gruyter, 2007. 595–600.

Erlich, Victor. *Russian Formalism.* 'sGravenhage: Mouton, 1955.

Eusterschulte, Anne. "Mimesis." *Historisches Wörterbuch der Rhetorik.* Vol. V. Eds. Gert Ueding et al. Darmstadt: Wissenschaftliche Buchgesellschaft, 2001. 1232–1294.

Forster, E.M. *Aspects of the Novel.* [1927] Harmondsworth: Penguin, 1974.

Gebauer, Gunter, and Christoph Wulf. *Mimesis. Kultur—Kunst—Gesellschaft.* Reinbek bei Hamburg: Rowohlt, 1992. (English translation: *Mimesis. Culture—Art—Society.* Trans. Don Reneau. Berkeley: University of California Press, 1996.)

Genette, Gérard. *Narrative Discourse. An Essay in Method.* Trans. Jane Lewin. Ithaca: Cornell University Press, 1983.

Gerrig, Richard. *Experiencing Narrative Worlds: On the Psychological Activities of Reading.* New Haven and London: Yale University Press, 1993.

Grass, Günter. *Günter Grass—Interview,* 1999. Web. 3 Oct. 2010. <http://nobelprize.org/nobel_prizes/literature/laureates/1999/grass-interview.html>

——. *The Tin Drum.* Trans. Breon Mitchell. Boston, New York: Houghton Mifflin, 2009. (*Die Blechtrommel.* 1959. Neuwied: Luchterhand.)

——. *Cat and Mouse.* Trans. Ralph Manheim. New York: Signet, 1964. (*Katz und Maus.* 1961. Neuwied: Luchterhand.)

——. *Dog Years.* Trans. Ralph Manheim. New York: Harvest, 1989. (*Hundejahre.* 1963. Neuwied: Luchterhand.)

Gross, Sabine. *Lese-Zeichen. Kognition, Medium und Materialität im Leseprozeß.* Darmstadt: Wissenschaftliche Buchgesellschaft, 1994.

——. "Narration in *The Tin Drum*: A Quirky Narrator in Search of the Truth." *Approaches to Teaching Grass's* The Tin Drum. Ed. Monika Shafi. New York: Modern Language Association, 2008. 75–89.

Hall, Katharina. "But Even Herr Matzerath Is Unable to Keep His Story Running in a Straight Line." *Approaches to Teaching Grass's* The Tin Drum. Ed. Monika Shafi. New York: Modern Language Association, 2008. 103–115.

Haycraft, Howard, ed. *The Art of the Mystery Story. A Collection of Critical Essays.* [1946]. New introduction by Robin Winks. New York: Carroll and Graf, 1992.

Herman, David, Manfred Jahn, and Marie-Laure Ryan, eds. *Routledge Encyclopedia of Narrative Theory.* London, New York: Routledge, 2006.

Hühn, Peter. "The Detective as Reader: Narrativity and Reading Concepts in Detective Fiction." *Modern Fiction Studies* 33:3 (1987): 451–466.

Jakobson, Roman. "Linguistics and Poetics." *Language in Literature.* Cambridge, Mass: Belknap/Harvard University Press, 1987: 62–94.

King, Laurie. *The Game.* New York: Bantam, 2004.

——. *Locked Rooms.* New York: Bantam, 2005.

Kleist, Heinrich. *Der zerbrochne Krug.* Stuttgart: Reclam, 1987. (English translation: "The Broken Jug." *Heinrich von Kleist. Five Plays.* Trans. Martin Greenberg. New Haven and London: Yale University Press, 1988.)

Kontje, Todd, ed. *A Companion to German Realism 1848–1900.* Rochester: Camden House, 2002.

Köppe, Tilman. "Kann man ein Buch spannend finden, obwohl man weiß, wie es endet? Überlegungen zum Paradox der Spannung." *Zwischen Text und Leser. Studien zu Begriff,*

Geschichte und Funktion literarischer Spannung. Eds. Ingo Irsigler, Christoph Jürgensen, and Daniela Langer. München: Edition Text und Kritik, 2008. 68–81.
Levine, Caroline. *The Serious Pleasures of Suspense: Victorian Realism and Narrative Doubt.* Charlottesville and London: The University of Virginia Press, 2003.
Mann, Thomas. *Der Zauberberg*. [1924] Frankfurt am Main: Fischer, 1967.
Martinez, Matias, and Michael Scheffel. *Einführung in die Erzähltheorie*. München: Beck, 1999.
Most, Glenn, and William Stowe, eds. *The Poetics of Murder. Detective Fiction and Literary Theory*. San Diego: Harcourt Brace Jovanovich, 1983.
Müller, Günter. "Erzählzeit und erzählte Zeit." *Festschrift für Paul Kluckhohn und Hermann Schneider*. Tübingen: Mohr, 1948 195–212.
Nünning, Ansgar. "Reliability." *Routledge Encyclopedia of Narrative Theory*. Eds. David Herman, et al. London, New York: Routledge, 2006. 495–497.
Poe, Edgar Allan. *The Selected Writings of Edgar Allan Poe*. Ed. G.R. Thompson. New York, London: Norton, 2004.
Priestman, Martin. *Detective Fiction and Literature. The Figure on the Carpet*. Basingstoke and London: MacMillan, 1990.
Raabe, Wilhelm. *Stopfkuchen. Sämtliche Werke*. (Braunschweiger Ausgabe, Vol. XVIII). Ed. Karl Hoppe. Göttingen: Vandenhoeck & Ruprecht, 1963.
Ronen, Ruth. *Possible Worlds in Literary Theory*. Cambridge: Cambridge University Press, 1994.
Ryan, Marie-Laure. 2006. "Possible-Worlds Theory." *Routledge Encyclopedia of Narrative Theory*. Eds. David Herman, et al. London, New York: Routledge, 2006. 446–450.
Sartre, Jean-Paul. *La Nausée*. Paris: Gallimard, 1947.
Sayers, Dorothy. "Aristotle on Detective Fiction." *Unpopular Opinions*. [1935] London: Viktor Gollancz, 1946. 178–190.
Shen, Dan. "Story-Discourse Distinction." *Routledge Encyclopedia of Narrative Theory*. Eds. David Herman, et al. London, New York: Routledge, 2006. 566–567.
Shklovsky, Viktor. *Theory of Prose*. Trans. Benjamin Sher. Elmwood Park, IL: Dalkey Archive Press, 1990. (*O teorii prozy*, Moscow, 1929)
Sophocles. *Oedipus the King*. Trans. David Grene. *Greek Tragedies,* Vol. 1. Eds. David Grene and Richmond Lattimore. Chicago: University of Chicago Press, 1960. 110–176.
Sternberg, Meir. *Expositional Modes and Temporal Ordering in Fiction*. Bloomington: Indiana University Press, 1978.
Storm, Theodor. *Der Schimmelreiter*. Stuttgart: Reclam, 1963.
Stowe, William. "From Semiotics to Hermeneutics: Modes of Detection in Doyle and Chandler." *The Poetics of Murder: Detective Fiction and Literary Theory*. Eds. Glenn Most and William Stowe. San Diego: Harcourt Brace Jovanovich, 1983. 366–383.
Vogt Jochen, ed. *Der Kriminalroman: Poetik—Theorie—Geschichte*. München: Fink, 1998.
Walser, Martin. *Ein fliehendes Pferd*. Frankfurt am Main: Suhrkamp, 1978. (English translation: "A Runaway Horse." *Three contemporary German novellas*. Ed. A. Leslie Willson. New York: Continuum, 2001.)
Wimsatt, W.K., and Monroe C. Beardsley. *The Verbal Icon*. Lexington: University of Kentucky Press, 1954.
Wolf, Christa. *Accident/A Day's News*. Trans. Heike Schwarzbauer and Rick Takvorian. New York: Farrar, Straus, Giroux, 1989 (*Störfall. Nachrichten eines Tages*. Berlin, Weimar: Aufbau, 1987.)

ORIGINS AS FUTURES IN THE TIME PLAYS OF J.B. PRIESTLEY

Carol Fischer

Abstract: J.B. Priestley (1894–1984), a prolific English writer, lived through two world wars and the twentieth-century's mix of physics and metaphysics, sociology and psychology, and called himself "a Time-haunted man." He used playwriting to illuminate certain theories about time that were circulating during his life and influenced his own ideas regarding time—especially precognition. Priestley was convinced that his culture's understanding of social justice, both its traditional perception and the possibility of change towards greater mutual care and equitability, was connected to people's perception of time. He also felt that the current state of drama in England should be adjusted to allow greater control by theatre artists themselves, so that stronger embodiment of his (and others') new ideas could more easily be presented. This paper addresses the place of Priestley's plays in dramatic history, and the insights that his theories and dramas offer towards the human experience of time.

Keywords: J.B. Priestley, P.D. Ouspensky, J.W. Dunne, time-haunted, eternal recurrence, déjà vu, serialism, precognition, time dimensions.

Imagine yourself being an audience to a play. The set on stage presents a quaint inn somewhere in the English countryside, June 1937. Staying at this inn are the main characters: Walter Ormund, age 45, his 27-year-old wife Janet, and a man around 35, Oliver Farrant, who was unacquainted with the Ormunds before their chance meeting in the country. Additionally there is a very odd, scholarly, older German gentleman who has a particular purpose in being at this particular inn and meeting these particular three guests. A professorial mathematician, Dr. Görtler is here to conduct an experiment, for he has had dreams and inklings about these three individuals. He is testing a hypothesis that his dreams are not only true life, but that his theory of "time recurrence" includes the possibility of intercepting a repeated life and shifting its future—basically changing the past.

The play is *I Have Been Here Before* (1937) by J.B. Priestley, a prolific English writer who published and produced plays in the middle of the twentieth century. Influenced by time theories circulating during his life, Priestley illuminated his contemplations and fascinations with the human experience of time in his playwriting; in this way he made these theories accessible to a larger audience, translating the more esoteric ideas into the type of entertainment that would both appeal to and rouse the average playgoer. In this essay, I explore his curiosity with time through analysis of two representative plays, *I Have Been Here Before* and *Time and the Conways* (1937), labeled "Time Plays" by the playwright himself in a 1938 published collection, *Three Time-Plays*. The third play in the collection, *Dangerous Corner* (1932), is looked at in less detail. The first two analyses include an overview of two time theories of the era. Additionally, I enlarge the discussion with some perspectives on other plays that use a "trick of time," as Priestley has described them. In all these plays, one observes that Priestley's overriding concerns are with the interaction of people within their society, humans caring for one another and consciously being together; he understood there to be a direct connection between how one views the nature of time and how one treats one's fellow citizens. Priestley's work was driven by theories that propose a nonlinear, extra-dimensional view of time, which allows an individual to peek into times other than the current moment, particularly into the future, and adjust his or her next step according to the insight gained. But the ability to steal a peek into one or more futures is not about gaining an advantage of decision or outcome so much as it is about situating an individual in a larger context of concern beyond the self. In short, glimpsed futures may guide us to decisions involving more agencies than our own ambitions and happiness. A corresponding logic for such understanding is central to the theory of time recurrence, the central thesis in the play introducing this essay.

In *I Have Been Here Before*, Dr. Görtler explains the idea to Walter Ormund, who is at a mid-life crisis point:

> We do not go round a circle. That is an illusion, just as the circling of the planets and stars is an illusion. We move along a spiral track. It is not quite the same journey from the cradle to the grave each time. Sometimes the differences are small, sometimes they are very important. We must set out each time on the same road but along that road we have a chance of adventures.

Ormund, who had been planning to commit suicide, responds:

> I wish I could believe that! ... They've never told me yet about a God so generous and noble and wise that he won't allow a few decisions that we make in our ignorance, haste and bewilderment to settle our fate forever. Why should this poor improvisation be our whole existence? Why should this great theatre of suns and moons and starlight have been created for the first pitiful charade we can contrive?

Görtler answers him:

> It was not. We must play our parts until the drama is perfect. (152)

In his depressed state of mind, Ormund cannot conceive that some god or other force in the universe would arrange things in such a way that he could have a chance through another life to make different decisions, but Görtler assures him that that is the case. To understand more of this theory, I first consider the playwright.

John Boynton Priestley was a journalist, essayist, novelist, as well as a playwright, with interests in social justice, cultural healing in the aftermath of the world wars, Jungian depth psychology including dream analysis, and the theories and experiences of Time. So important to him was this latter interest that he wrote an extensive interdisciplinary study published in 1964 entitled *Man and Time*. The original publishers, Aldus Books of London, thought the book significant enough that they printed the volume as a companion to Carl Jung's *Man and His Symbols*. Although Priestley called the tome a "personal essay" because he treats the material from a highly subjective viewpoint, the book is a beautifully organized, 300-page work that resembles a coffee-table volume with a wealth of photographs and references.

The book's opening title and acknowledgement pages are framed with illustrations of the months of the year from a medieval prayer book. Part One is a history lesson. Priestley records information about various kinds of sundials from the ancients to medieval hand-dials, from Stonehenge to Egypt, water clocks, mercury clocks, and hour glasses. He leads the reader through discussions of pendulum clocks and calendars, the arbitrary time division of the week, and other measurements of time. He continues with various artistic images of time, psychological anomalies of perceived time, and the metaphysics of John McTaggart. A quick trip through the historical science of time up to and encompassing related possible theories in quantum physics follows before Priestley touches on treatments in fiction and drama, including that of H.G. Wells, Samuel Beckett, his own, and the works of other playwrights. The discussion moves into a psychological view of humankind's experience of time from "The Great Time" of ancient mythologies forward to the present age. Only then does he move into the theories that possessed him as he wrote the plays *I Have Been Here Before* and *Time and the Conways*.

It is in *Man and Time* that we learn of his extraordinary thoughts regarding precognition and dreams, and peruse the information and examples he broadly gathered in support of his ideas. He relates that during a television interview on the BBC program *Monitor* in 1963, Priestley asked viewers to send him "accounts of any experiences they had had that appeared to challenge the conventional and 'common-sense' idea of Time" (187). To the hundreds of replies, he added his own dream experiences and wrote at length to corroborate proof of precognition, of glimpsing what will happen in the supposed future. He called himself "a Time-haunted man addressing himself chiefly to all those people he knows from experience to be also Time-haunted" (12). Priestley also introduces the theory of time recurrence as understood by P.D. Ouspensky, the chief disciple of G.I. Gurdjieff, a teacher of esoteric thought and psychological development.

This essay's introductory dramatic moment was brought to you by Priestley's contemplation of Ouspensky's assertion of eternal recurrence. Providing a proof text for Jung's theories about synchronicity, the playwright in his autobiographical *Midnight on the Desert* (1937) narrates the story of his discovery that Ouspensky's work expounded this unique time concept. While on a trip to California, Priestley spent some days at the old Biltmore Hotel in Santa Barbara, a favorite picturesque resort on the beach. He wandered through a section of Santa Barbara called "El Paseo," an old outdoor downtown mall with quaint restaurants and shops. (He would not recognize the revamped side-street today with its law offices and galleries). There he discovered a small bookshop, entered, and began perusing shelves. Priestley recounts, "a minor miracle happened" (243). A book by Ouspensky he had seen referred to several times while researching time, the book he had wanted to find but could never locate, was sitting on a low shelf as if patiently waiting for him. This book was Ouspensky's *A New Model of the Universe.* (The book's first translated edition was published in 1931, but the preface indicates that most of the book was written in Russian before 1914 and finished around 1919. Priestley's autobiographical story of this finding is narrated in *Midnight on the Desert,* 242–244.)

As a maturing occultist in the early decades of the twentieth century, Ouspensky promoted new models of esoteric thought. His writings include discussions of physics, Christianity, the tarot, yoga, dreams and hypnotism, the idea of superman, mysticism and magic, the miraculous, sex and evolution, and time as an order of eternal recurrence. He asserted that solutions to the fundamental problems of birth and death would directly follow on one's understanding of time. If one believes time is a line that comes into existence at birth and ceases at death, then everything outside that line is by default unknown and unknowable. But the deeper mystery of time is actually in the existence before and after that limited, finite line. So in Chapter XI of his book, Ouspensky develops the idea of eternal recurrence, related to reincarnation, but different in essence. He presents our life cycles as being circles within waves. The point of death on the circle is also the point of birth even as the wave is moving in a direction called the future: "One time ends and another begins. Death is really a return to the beginning" (477). And the exact same life recurs ad infinitum. Therefore, the sensation of déjà vu, one's experience of "this has happened before," is because the moment indeed *HAS* happened before. It is the absoluteness of repetition that creates the feeling of fate, of inevitability, that people either accept or rail against. I think, too, that the idea subtly permeates the historically defined idea in Western culture that individuals act out roles in their lives. For instance, both theatre and painting frequently treat the analogy of life as a stage and humans as players. Each person fits into a slot or role already prepared for him or her; for Ouspensky, this is not because some god has ordained or predestined that role, but because s/he has already played that role, not just once but multiple times. In the individual consciousness behind this concept, however, is a different philosophy regarding the meaning of one's life. A person who is trying to act according to "God's plan" would have a forward thrust and purpose; anxiety, consternation, and struggle come when one is misaligned with the divine path. But Ouspensky describes a more humanistic based journey with many pitfalls.

Ouspensky admits there are "tendencies" in the preceding lives that strengthen for the better or degrade for the worse in succeeding lives. The increasing tendency will either raise or lower the vitality within an individual. Those with manifestations toward degeneration such as "failures, drunkards, criminals, prostitutes, and suicides" experience their life fiascos more and more readily with each repetition, which weakens their vital force until they pass slowly out of life altogether, finally ceasing to be (re)born (482–483). He regards as fact that the soul is finite and also fades from existence in such a case. In contrast, yet parallel to those poor unfortunates, are others who, in repeating lifetimes, find success more quickly and effortlessly with each go-round, but nonetheless are in danger of a growing mechanical aspect in their behavior that saps their life vitality. Such people will begin to assume that all their success arises fully from their own merit, which deadens other soul-enhancing possibilities: "They sacrifice everything to success ... and they suppress in themselves all other desires, interests and inclinations" (484).[1] (Perhaps Tiger Woods would have benefitted from such knowledge.)

The most favorable disposition is in "men of real science, of real art, of real thought or action" (484). In each life cycle they aim for the unattainable ultimate good with new strength and inner development. Now, it is understood that the struggle against evil is a struggle with the past, for the causes of evil are in the past. There can be no evolution if it is not possible for such favorable men to move into the past to fight evil and to alter things. This "travel" to the past is possible if time itself has three dimensions similarly to space. In the way that space is understood to fill dimensions one, two, three, Ouspensky understands that time occupies dimensions four, five, and six. The fourth dimension is an extension of time usually understood as the third dimension in motion, but is also a closed circle beginning with the moment of birth and ending at the same point at the moment of death (460). The fifth dimension is a further extension or eternity, which Ouspensky contemplated as time repeating itself (477). The sixth dimension is "the stretching out of these eternal circles into a spiral or a cylinder with a screw-thread in which each circle ... passes into another circle" (461).[2] An individual who has attained great consciousness in his/her evolutionary journey recollects (although vaguely) previous lives, and escaping into the spiral of the sixth dimension, returns to a previous life cycle, but with the advantage of knowing the results. From one's own conscious vantage point, one can gently influence people's thoughts, decisions, and actions towards more beneficial consequences in the future, both for individuals and for the larger society (483–486).[3]

Priestley demonstrates these temporal dynamics in his play. Although appreciating Ouspensky's efforts, he did not agree with eternal recurrence, and in *Man and Time* he takes issue with the Russian's offering unscientific, experiential statements as fact. Nevertheless, Priestley uses the time

---

[1] Ouspensky includes two additional categories (481–482): those who never change but live a "petrified" routine life, and those whose lives are linked with great cycles such as reformers, revolutionaries, conquerors, emperors—their destinies are tied to the destinies of nations for good or ill, and therefore recurring lives cannot be different.

[2] Because Ouspensky approaches these ideas through esoteric thought and not Newtonian scientific experiment, these concepts regarding multiple dimensions seem convoluted and like self-fulfilling prophecy. For his purposes, however, they prove his idea of eternal recurrence.

[3] Ouspensky believes that true evolution, i.e. inner development, is only available through intense study of a special doctrine accessible in certain types of yoga.

concept in *I Have Been Here Before*, softening it through a connection to a more traditional understanding of God as illustrated by Ormund's response to Görtler quoted earlier. The professor has retained some conscious knowledge of the future through his dreams, and he knows there is great misery in store for all players after Ormund shoots himself at the country inn. The mathematician explains what he knows of their lives to the three others. Ormund and Farrant at first dismiss it all as the ravings of a mad old man, but the idea makes sense to Janet who had been acutely aware of the sorrowful atmosphere, the sense of tragic déjà vu, haunting the weekend. Eventually, with new acknowledgement, new consciousness, Ormund retreats from suicide, choosing instead to release his wife to go forth with her new love, the man Farrant, whom she had just met this weekend but with whom she had experienced immediate attraction, *as if they had known each other a long time*. If they had secretly run away together as had happened in previous life cycles, Ormund would have shot himself, the new couple would have been disgraced and destitute, and others who had invested in Ormund's business would have suffered financial ruin. Görtler embraces what the new choice means for Ormund. He tells him, "you are now in the unusual and interesting position of a man who is moving out on a new time track, like a man who is suddenly born into a strange new world—It may be hard at first. But…there are a million suns waiting to keep you warm and to light your way" (155). Although sounding like a bit of schmaltz to cynical ears, the line sets nicely after the play's shadowy foreboding and Ormund's self-deprecation and hopelessness.

Priestley's knowledge of what makes for good theatre, added to his expertise in playwriting, created a drama that is more than the exposition of a bizarre theory of time; he created a work that translated well into visual staging and action, and generally pleased his audiences with vivid characters and captivating dialogue. From a young age Priestley enjoyed the entertainment of theatrical performance, whether in music halls and variety shows, or touring plays. He appreciated the live-ness of the art, the interchange of audience and actor, the "now"-ness which he felt could not be replaced by film or television, even though those latter forms were growing in popularity (Gale 27–28). Driving his theatre work was the expectation "that theatre had the power to educate and to integrate and enliven a community, to help it to think through the social issues which affected it" (35). What he did not like was the economics of the theatre industry that developed from the time of World War I into the 1930s, when those with adequate financial resources owned and controlled the management of theatres (and by implication also the product) rather than the dramatists, actors, and producers creating the art. This dissatisfaction compelled him to write often about theatre in newspapers and journals, and to start his own small production company. He envisioned enchanting an audience with all the tricks that one can perform in the magic of theatre, and considered his time plays to reflect that magical style (38). Gareth Lloyd Evans, whose book focuses on Priestley as dramatist, perceives the playwright's concern being "to convince that the magic and the mystery swirls [sic] about us, that to be aware of it is to be aware of the oneness of humanity" (147). Written with a keen, dramatic sense of storytelling connecting human beings to others and to themselves, Priestley's plays continue to draw interest from today's audiences as well as producers. Indeed, what sparked my interest in Priestley and his plays was a 2009 production of *Time and the Conways* at the University of California, Santa Barbara.

Time and the Conways is one of Priestley's most often produced plays. Some early critics felt that he made a mistake in setting the three acts out of chronological order. Acts One and Three take place in 1919, while Act Two, in between, takes place in 1937, eighteen years in the future, reflecting the same year as the first production. The play begins and ends on the night of a party celebrating the twenty-first birthday of the Conway daughter Kay, an aspiring novelist. Completing the family circle is a rather silly widow mother, one older and two younger sisters, a "slow," perhaps today we would say "geeky," brother, and a brother just returned from the war, who is not nearly as perfect as his mother imagines. Act One is full of fun. We see the preparations for a charade-type game, a dramatic device that introduces the unique traits of each character and that elicits much excited anticipation about the future, including weddings and successful careers. Even the political consciousness in one of the sisters exudes optimism about post-war possibilities and socialism as a political panacea.

In writing the play, however, Priestley deliberately chose to illustrate J.W. Dunne's time theory "Serialism," while also commenting on socio-political realities and exploring the workings of a family in a Chekhovian style with the characters' sometimes distracted dialogue, the bantering and conflicts, the hopes and dreams, and the family members' naturalistic messes. Dunne was an aeronautics engineer concentrating on design and stability of craft, including engineering World War I planes. His secondary, but passionate, interests lay in parapsychology, dreams, and the nature of time. He developed the concept of Serialism with infinite regress, that is, with no limitation or natural stopping point. The theory presents an individual as a so-called "first observer"; s/he is a given with a definite physical beginning, not an entity deduced from logic or abstract thought as are successive "observers." This individual soul "has a definite beginning in absolute time—a soul whose *immortality, being in other dimensions of Time, does not clash with the obvious ending of the individual in the physiologist's Time dimension*" (120; italics in the original). To Dunne that meant that there is a "superlative general observer" containing this individual soul but who works towards differentiation from all the other souls contained in that superlative position (120).

Priestley considered Dunne primary in removing the speculations of time from the arenas of philosophers and mystics to a scientific yet experiential realm within quotidian existence. In *Man and Time* Priestley writes about Dunne: "He rejects the idea… that our lives are completely contained by chronological uni-dimensional time," and he does so without becoming other-worldly (245). Dunne then "spatializes" time into multi-dimensions (similarly to Ouspensky, but with more graphs explaining the physics). A second time dimension includes a second observer timing the first dimension, but then another observer/dimension must account for the second, etc.[4] He draws this out from the dream dimensions of an individual, a place where the borders between past, present, and future are insubstantial and the dreaming observer can step across into the future as well as the past, so that the dreamer, when conscious, experiences precognition of events. The dreamer dwells within another level of consciousness or dimension as an observer of the first; then there is a third watching the second watching the first, and so on. But there is more. Observer two is able to watch the future of observer one through precognitive dreams, and can intervene in observer one's present time consciousness so that a shift in observer one's actions at a particular moment changes what initially would have been the future state of affairs.[5]

Priestley affirmed and negated some of Dunne's thinking. He agreed that the first time dimension is an individual's experience—the person

[4] One is reminded of physicist Eugene Wigner's thought experiment about observers in quantum physics who are integral to actualization from possibilities. A second observer would treat a first observer as part of the experiment and not necessarily as a conscious observer. To resolve the dilemma, it is agreed that the first conscious observer is the only one integral to the result.

[5] Priestley offers a lengthy interpretation of this concept of changing the future through a woman's recounting a dream where her toddler had drowned. Weeks later, when the woman found herself in the midst of experiencing her dream situation, she shifted her actions, and the baby did not drown (*Man and Time*, 258–259).

as observer one. The second is where there is an observer who can watch the individual—a dimension of more *ultimate* time. And the third time dimension is an absolute time that includes all past, present, and future for an additional observer. He does not regress to acknowledge possible further innumerable dimensions. He does, however, posit that all creative endeavors originate within this third time dimension. In dream time, when observer one is not focused on the first dimension, an individual can access, as observer two and/or three, one of the other dimensions crossing past, present, and future.

More important to Priestley than numbers of dimensions or observers was the dreamer's ability to access the future in absolute time—the fact of precognition. This is what the playwright illustrates in *Time and the Conways*. The end of Act One puts much stress on the actress playing Kay who must silently express an internal shift in such a way that the audience can read her descent into a dreamlike state. The stage directions read:

> *She quietly switches off the lights inside the room. She draws open the curtains at the window and sits down on the window-seat, looking off left, listening to the song in a sort of ecstatic reverie, quite still. As Kay listens, without moving a finger, staring away from the exit, not at something but into something, and the song goes on, the curtain slowly and quietly descends.* (32)

Act Two is much more somber in tone. The reality of crushed dreams has taken a heavy toll on the family members. The act's timeframe, reflecting the time in which Priestley was literally writing the play, is 1937; World War II moves towards inevitability both in the drama and in reality. Our dreamer, Kay, has discarded the idealism of writing only what is meaningful to her and works as a journalist. (One wonders if Priestley is speaking of his own experience as a writer.) The politically astute sister is a single, embittered schoolteacher; the youngest sister has died; the returned soldier brother has abandoned wife and children to poverty and drowns his own sorrow and vagrancy in drink. The oldest sister is in an emotionally if not physically abusive marriage, and the mother has to sell the house at a loss as she has squandered all her inheritance on her favorite, albeit drunk, son. Kay and the geeky oldest brother Alan, the most stable and content of the lot, have a conversation in which Alan defines a belief about time that echoes Dunne.

> Alan explains: Time's only a kind of dream. If it wasn't, it would have to destroy everything—the whole universe—and then remake it again every tenth of a second. But time doesn't destroy anything. It merely moves us on—in this life—from one peephole to the next.... We're seeing another bit of the view—a bad bit, if you like—but the whole landscape is still

> there... What we *really* are is the whole stretch of ourselves, all our time, and when we come to the end of this life, all those selves, all our time, will be *us*—the real you, the real me. And then perhaps we'll find ourselves in another time, which is only another kind of dream.... Take a long view.
>
> Kay asks: As if we're—immortal beings?
>
> Alan responds: Yes, and in for a tremendous adventure. (59–60)[6]

Since this conversation between Kay and Alan occurs at the end of Act Two, we continue in Act Three by returning to the 1919 dramatic present, but the family banter has lost any charm for the audience because they now know the demise ahead. Kay is nervous and disoriented because of the glimpse she had of their depressing future. She cannot recall details, only residual feelings and vague possibilities. The sentimentality in the family's conversation sounds frightening and absurd to her rather than sweetly longing and hopeful. There is bitterness or cynicism or perhaps horror in the closing scene as Carol, the youngest sister, exclaims: "The point is—to live. Never mind about money and positions and husbands with titles and rubbish—I'm going to live" (80). Because of the time leap in Act Two, we know that Carol will die within a few years. Kay begs Alan to tell her something—from the memory of her experience of the future there is something that he had said to her—that would offer comfort. Alan's response seems to indicate that he shares her precognition: "There will be—something—I can tell you—one day. I'll try—to be wise—I promise" (82). The terror of the future recedes but without fully dissipating as the lights fade and the curtain falls. Understanding that the time concepts of regress and multi-dimensionality open certain participants to precognition is central to interpreting a meaning well beyond fantasy. There is the recognition that even though World War I should have been "the war to end all wars," the country was slipping towards war again and few seemed to be paying attention. Kay stands in for the audience as she faces the dilemma of knowing something about what is ahead, but is unsure how to shift the path or even if such a shift is possible. She suffers the psychological experience of the haunting nature of the human condition: how much control over our own individual, family, and community lives do we hold? A wise individual like Görtler is sorely needed, or the

6 A curious digression here is the echo in Alan's line to Peter Pan's line: "to die will be an awfully big adventure," a quote well known from J.M. Barrie's play *Peter Pan*. Barrie, a personal friend of Priestley, wrote plays until he died in 1937. In *Man and Time*, Priestley recalls that Barrie's plays "would reveal many strange elements—one of them... involving Time" (134).

intervention of an observer two into the consciousness of an observer one whose actions could shift the future from the one seen in the precognitive, dreamlike vision.

In his own estimation, Priestley's success as a produced playwright is attributed more to his grasp of drama, character, story, and audience than to his being able to translate theories of Time into stage-worthy productions. He writes in *Man and Time*:

> A novel or a play cannot be *about* Time.... Time is a concept, a certain condition of experience, a mode of perception, and so forth; and a novel or a play, to be worth calling one, cannot really be about Time but only about the people and things that appear to be *in* Time. Some novelists and dramatists may be unusually aware of Time, but they have to write about something else. (122)

What I see as his focus in these time plays is an exploration of how his characters respond to the time dilemmas in which Priestley has placed them. Until thinking about it after the lights are up, an audience might not even realize that the crux of the play relied on the insertion of a trick of time. Fascinating, too, is that he seems not to repeat any particular dilemma among his several plays involving time, which could indicate that Priestley never completely settled for himself the question of the nature of time during his playwriting years. The plays are not variations on one theme, but a series of meditations on human effort within the unique space-times in which the characters find themselves. His first "time play," and first effort at playwriting does not involve as complex a theory as either Ouspensky's or Dunne's, but nevertheless illustrates his interest in involving characters in anomalies of time.

Dangerous Corner, produced in 1932, initiated Priestley into the company of playwrights. He explains in an introduction: "I constructed it to prove I could think and create like a dramatist and not necessarily like a novelist, and also to make use of the device of splitting time into two thus showing what might have happened, an idea that has always fascinated me" (*The Plays of J.B. Priestley* viii). The play begins at a dinner party with English-style bantering and playfulness bordering on spite, leading into a conversation about the merits of truth telling, half-truth telling, fibbing, and outright lying.[7] A musical cigarette box is opened and passed around

[7] Several of Priestley's plays either start with or include dinner parties—a dramatic convention that he uses to great effect to listen in on conversations that reveal much about those involved and provide an audience with necessary exposition.

with its offerings, but this simple box is a Pandora's box for this group of men and women because it belonged to brother-in-law Martin who had died in mysterious circumstances the previous year. The party quickly disintegrates as Olwen acknowledges that Martin had shown him exactly this cigarette box, but backs away from this observation when Freda insists that Martin did not own it when Olwen last visited him. Freda's husband Robert pursues the matter. "Olwen, I'm going to be rather rude, but I know you won't mind. You know you suddenly stopped telling the truth then, didn't you? You're absolutely positive that this is the box Martin showed you, just as Freda is equally positive it isn't" (22). The truths about the relationships among the dinner party guests and about Martin's death are slowly and painfully brought to light in a manner that reflects an earlier comment: "Telling the truth is about as healthy as skidding round a corner at sixty" (20). At the end of the destruction, Robert is on his way to suicide: "I'm through. There can't be a tomorrow," he exclaims just before exiting to his room where he will procure a pistol. Then a woman screams: "Stop, Robert! Stop!" (83).

What stops is the entire play. And then what begins is the entire play. The next lines are almost an exact repeat of the opening. In this round, however, when Freda opens the cigarette box, instead of Robert's accusation and the round of searching, destructive questions accompanying the smokes, music on the radio drowns out the music of the box. Chairs are pushed back, and everyone begins to dance. The "truth," at least for the time being, has been avoided. Priestley returned to the moment before Time split and envisions another option as he closes the play. Until the point of re-embarkation, we as audience are not aware that what we have been watching is merely one possible timely situation. And we are also not privy to the outcome of Robert's probable suicide. The last lines are what might have happened if this staged world had proceeded in a direction other than, but existing simultaneously, with the first. With the return to the beginning, the guests are offered a chance to change the future, although crucially they are not aware of that chance. Priestley treats that moment of chance as a time split rather than a happenstance, or a precognition, creating a second future from one original moment.

Verisimilitude in a theatrical production captures a particular audience's consciousness of contemporaneous cultural interests and philosophies. The different pictures shaped by the pressure of time theories in Priestley's plays are comparable to theories that physicists have proposed about the nature of time not only during Priestley's own life but also in our contemporary time. *Scientific American* published a special edition,

A Matter of Time, Spring 2012, that details some of the current complementary and conflicting approaches to the question of time. One suggestion by philosopher Craig Callender (in an article considering anew the question of whether time is an illusion) illustrates how he considers that "time is not like space." Callender encourages the reader to imagine slicing a box containing a bouncing ball in one direction (say west to east so that slices are next to each other like a loaf of bread); such slicing would reveal a sequence of the ball's progression at a diagonal from bottom to top or top to bottom. This is equivalent to the standard way of slicing a box of space-time with time moving from past to future. However, if the box is sliced in the other direction (north to south so that slices are on top of one another like a multi-layered sandwich), the ball would be seen in succeeding slices in more random positions with sometimes two balls in one slice as the ball had passed through that slice on its way up and its way down (17). Rather than being able to use the slices to predict where the ball will be at some future point as one can with the first series of slices, the observer is looking at a complete event of past, present, and future, and must use all the slices to understand the picture. With the second slicing, Callendar suggests the unification of time that Priestley explores through notions of precognition and the alternative consciousness of dreamtime. All the plays together do not add up to a sequential thought process reaching towards definitive conclusions about either the nature of time or of humanity, but through his plays Priestley ponders what would be the various slices with the goal of impacting the viewers' awareness of connections to themselves.

The dramatic details of those ponderings pointedly vary in his other plays involving time interacting with humans as well as in the three plays I have already considered. The writer points to his own interest in discovering new perceptions through playwriting in his notes on the play *Johnson Over Jordan* (1939): "What I wanted to do was to give an account, in dramatic form, of a man's life in a new way, taking an ordinary middle class citizen of our time and then throwing a new light on him and his affairs, and giving to my record of his rather commonplace life an unusual depth and poignancy" (120). *Johnson Over Jordan* imagines a man's afterlife through his odd revisiting of settings and events significant to his life through time travel to his pre-death past. The Everyman character Johnson does more than merely remember his worldly existence; he moves "freely in and out of the past and recreates the experiences he finds there... [he] is back in the past and yet fundamentally out of time altogether" (125–126). This may seem like "old hat" to us who have seen many creative post-death and dream interpretations through the magic of film (*Inception* and *What Dreams May Come* immediately come to mind). But Priestley produced *Johnson Over Jordan* as a theatrical experiment for a London theatre scene that he considered "about the least experimental in the world, and always badly in need of a production that is not exactly like the last two hundred and one productions" (133). He added his lengthy notes to the published play because many critics and attendees had expressed puzzlement and complained about a lack of form and meaning in the play. In short, he felt compelled to justify the work. His concern for the state of England, its theatre, and the emotional state of the people in 1939 is clear: "If we could take their minds away from their own immediate concerns and yet at the same time make them feel deeply in the playhouse, provide them with a searching experience, then we might do them all a power of good, and at the same time clear a way through the thickening jungle of mass-produced trivialities for bigger and nobler work than ours" (134–135).

Two other Priestley plays use different styles of perceptions and imaginings in time to encourage people to feel deeply and leave the theatre in a searching frame of mind. In *Music at Night* (1938), while a pianist is playing at a private party, the relatives and guests create varying scenes in their imagination that are actualized on stage. As in Ouspensky's idea of the sixth dimension, observer three in absolute time takes over for observer one who is absorbed in the music, and other possible lives are articulated.[8] Glimpses into alternative lives and responses are exposed in those moments: past lives, other present lives, and future lives. One elderly man even dies and proceeds into his own other life as the play ends. Written around the same time as *Johnson Over Jordan*, *Music at Night* features the lives outside of life and creates a strong sense of need for people to search deeper into their own psyches in order to understand who they are and how they relate to others. *They Came to a City* (written at the war's end in 1945) illuminates a time-outside-of-time equivalent to post-trauma unconsciousness from which a victim will either die or return to vibrancy. The various characters, who have all recently suffered accidents that they cannot quite remember, are not sure where they are, but are offered the choice to move into a nearby beautiful, socially just, utopian city whose

8 Although two decades before the physicist Hugh Everett published work (in 1957) on his Many Worlds Interpretation of quantum mechanics, the idea of simultaneously existing lives bears close resemblance to that sub-atomic theory.

door is open for one day, or return to their known world of bigotry, money, and social class distinction. While several characters return because they hate the lack of social class and the futility of money in the shining city, one couple chooses to return in order to proclaim to a morally impoverished world the possibility of creating a satisfying community based on social justice. Each of these plays is a vehicle for Priestley as he tests various ideas about time and its influence on human philosophies and relationships. This second play is especially poignant when one considers it was written at the close of World War II, a time before things could return to pre-war status quo, and thus a time open for social change.

The last play I discuss in this essay is the best known of Priestley's plays, *An Inspector Calls,* written and first produced in 1945 (the same year as *They Came to a City*), made into a film in 1954, a television miniseries in 1982, produced on BBC radio, and revived on stage in England and New York in the early 1990s, receiving the 1994 Tony award for the Best Play Revival. The play begins with characters in evening dress seated around a dining table, a convention Priestley used to draw in England's conventional audience, accustomed to a naturalistic theatrical style, to make them comfortable enough to hear the message. The participants are preparing to enjoy port, cigars, and cigarettes after a sumptuous dinner. After a round of toasting to congratulate the hosts' daughter's engagement, the women customarily withdraw, and the men continue a banal conversation regarding social status and the rumors of war. The time is set in spring, 1912.

The niceties are interrupted by the doorbell and the entrance of Police Inspector Goole. It seems that a young woman named Eva Smith has committed suicide by drinking some poisonous disinfectant that horribly burned her insides; the Inspector wants to ask questions of the Birling family and the daughter's fiancé. At first there seems to be a mistake, for how could anyone in this well-off social gathering be involved with a destitute woman from a lower social class. But question by question uncovers that each person at the dinner party had a part in the pregnant young woman's fall into poverty and desperation.

When a group of girls under his employ had asked for a small increase in wages two years ago, Mr. Birling had fired all of them, including Eva Smith. Birling's daughter Sheila had been in a temper one day when shopping in the clothing store where Eva had subsequently found work. Eva unwittingly irritated Sheila, who then told the owners to fire her or the Birlings would close their account. In the summer previous to this evening, Sheila's fiancé Gerald had "taken up" with Eva (who had been using the name Daisy), but had broken it off after a few months when he decided to pursue Sheila again, this time more wholeheartedly. Mrs. Birling denied charity to Eva when she approached a committee seeking aid because she was pregnant and had no money. Indeed, Mrs. Birling had told her to find the child's father and make him take responsibility. Eric Birling, the son, finally admits to having fathered the baby, and to stealing money from his father's company to support Eva, who refused further help after discovering its illegal source. After all present are compelled to admit their complicity in the girl's death, Goole chastises them with an intense monologue regarding social responsibility, and leaves.

The first reaction after Goole leaves is to lay blame on someone. Mr. Birling chooses Eric because of Eva's pregnancy, but is more concerned about a possible public scandal than he is about the damage to the girl. Sheila is the one who seems most able to accept the stark facts, including her own part in Eva's demise. Eric points out his father's culpability in his talk with the men after dinner. Birling had told the younger two that a man must "look after himself.... the way some of these cranks talk and write now, you'd think everybody has to look after everybody else, as if we were all mixed up together like bees in a hive—community and all that nonsense" (168). Since it was on the instant that he put a period on that sentence that Goole rang the doorbell, Sheila and the others begin to wonder whether the Inspector was actually an official police inspector or merely an imposter, a hoaxer. After all, he seemed to know everything already, and he never showed Eva's photo to two people at the same time, meaning he could have shown them all different photos. Gerald returns from walking (having left after he had been interrogated), relates that he had asked a police sergeant on the street about Goole, and discovered that no one by that name or description is on the force.

The recent confessions turn to anger as the thought grows that they have been hoodwinked, and Birling is sure if they all stay quiet, there will be no public knowledge to worry about. Perhaps the girl who died had nothing to do with any of them. Sheila, however, stands by the guilt of their shameful actions no matter who died from the poisoning. Gerald calls the hospital to verify that some girl had indeed died and discovers that no suicide had been brought in today or any day for months. All but Sheila, who still considers them guilty but only lucky in the outcome, begin to relax. As they consider themselves in the clear, the phone rings "sharply," and it is the police saying that an inspector is on his way over to talk to them because a girl has just died on her way to the infirmary after swallowing disinfectant. The shivering sensation left is that Goole had time travelled through all their times and knew all their secrets. He was a third observer in absolute time to each of their first observers.

The impact of the play, however, is not in its surprise conclusion seemingly crossing from one time dimension to another, but in its stance on social justice and community consciousness. Priestley makes a statement about the lack of responsibility of a certain social class for the majority of society and that such a state of blindness in the wake of World War II was unacceptable. Just before Goole's leave taking the "Inspector" warns them:

> There are millions and millions and millions of Eva Smiths and John Smiths still left with us, with their lives, their hopes and fears, their suffering, and chance of happiness, all intertwined with our lives, with what we think and say and do. We don't live alone. We are members of one body. We are responsible for each other. And I tell you that the time will soon come when, if men will not learn that lesson, then they will be taught it in fire and blood and anguish. (207)

Priestley emphasizes the significance of that statement by setting the play just before World War I. His audience certainly had cultural memories of "fire and blood and anguish," both recent and only a few decades old that would be pricked by those words. The intensity of Goole's monologue creates a stronger and more blatant message to the audience than even the socially just utopia in *They Came to a City*.

Priestley was a social activist and commentator as well as a time-haunted playwright, and *An Inspector Calls* overtly exposes him on all points. John Baxendale, in his recent book, *Priestley's England*, while not focusing on the element of Time, notes that the play "is not a plea for an agenda—a set of political proposals—but for something deeper: an ambience—'the total climate of values, ideas, opinions, fears and hopes, in which we live' " (169). Priestley experimented in his plays using dramatic devices focused on time to encourage people to see their lives from a different point of view, to become sensitively conscious of the world inside and around themselves. Using dramatic entertainment, these many plays placed the embodied choices of the mid-century British culture *and* the alternatives in an audience's direct line of vision, even while Priestley challenged the conventional London theatre to experiment and think more deeply about the potential of its audience at the time. Baxendale ends his treatise on Priestley with a paraphrase from one of the playwright's later critical works, *Literature and Western Man* (1960): "[If] we try to establish justice, order, and real community in the outer world, and accepting the great mystery of existence, openly acknowledge the deepest needs of the inner world, we yet find the right symbolic tools to restore our wholeness" (188). Priestley concludes *Man and Time* with a final connection to our origins (as choices for new action) and futures: "I have written this book in the belief that the choice of the right turning, a decision that may be final if not for our whole species then at least for our civilization, cannot be separated from the relations between Man and Time" (308).

I see a remarkably beautiful link between J. B. Priestley and the thoughts of ISST founder, J.T. Fraser. Priestley would agree with Fraser's statement in his introductory address at the 2008 conference "Times and Worlds" in Mexico City, "that human values—the true, the good and the beautiful—are made possible by the human sense of time. Also, that their pragmatic, long-term roles, are to keep humans sufficiently challenged to remain creative, but also sufficiently tamed, not to exterminate themselves" (11). These two authors have much to say to one another and to the world. What Priestley worked towards in his writings, broadcasts, productions, and theatrical reinventions, Fraser brought to fruition within his own sociological, philosophical, and scientific contributions. Both were "time-haunted" individuals. For the compelling concept of humanity's social unity for which both authors urge us to strive, and in line with the needs of communities and global culture today, Priestley's Time Plays are indeed still "timely."

References Cited

Baxendale, John. *Priestley's England*. Manchester: Manchester University Press, 2007.

Callender, Craig. "Is Time an Illusion?" *Scientific American* 21.1 (2012): 15–21.

Dunne, J.W. *An Experiment with Time*. Charlottesville, VA: Hampton Roads Publishing Company, 2001.

Evans, Gareth Lloyd. *J.B. Priestley—The Dramatist*. London: William Heinemann, 1964.

Fraser, J.T. "Whose Past Is Our Prologue?" *Kronoscope* 9.1–2 (2009): 5–12.

Gale, Maggie B. *J.B. Priestley*. New York: Routledge, 2008.

Ouspensky, Peter. D. *A New Model of the Universe*. Mineola, NY: Dover Publications, 1997.

Priestley, J.B. *An Inspector Calls and Other Plays*. London: Penquin Books, 2000.

——. *Johnson Over Jordan*. London: William Heinemann, 1939.

——. *Man and Time*. London: Bloomsbury Books, 1989.

——. *Margin Released*. New York: Harper and Row, 1962.

——. *Midnight on the Desert, A Chapter of Autobiography*. London: William Heinemann, 1937.

——. *The Plays of J. B. Priestley, Volume One*. Kingswood, Surrey: Windmill Press, 1962.

——. *They Came to a City*. London: Samuel French, 1944.

——. *Three Time-Plays*. France: Pan Books, 1947.

——. *Time and the Conways*. London: Samuel French, 1967.

BIG SCIENCE: MARCHING FORWARD TO THE PAST

Michael Crawford

> We shall not cease from exploration
> And the end of all our exploring
> Will be to arrive where we started
> And know the place for the first time.
>
> T.S. Eliot, *Four Quartets*

Abstract: Although there have been several large scale science projects over the past century, the last two decades have seen the first mega-projects that have emphasized *basic* as distinct from *applied* science. Inevitably, in an effort to sell and justify the large cost of these recent multi-national efforts, scientists have taken care to highlight the huge future conceptual, practical, and economic payoffs that are likely to accrue. Despite this ostensible orientation towards the future, there is a recent and remarkable concordance of interest in origins evident in recent multi-billion dollar basic science projects. Even those projects that have an explicitly future-oriented agenda contain within themselves the seeds of a profoundly retrospective obsession. I will outline how an established undertaking, the Human Genome Project, was originally pitched to spotlight future applicability, and how it ultimately came to focus upon origins and evolution. I will then show how a recently developing subsidiary consortium, the Human Epigenome Project, contains similar seeds of retrospection even as it develops an explicitly forward-oriented agenda. Finally, I will illustrate how a likely future (and futuristic) "Big Science" project, Exobiology, is actually starting off by looking at the evolutionary origins of life on Earth.

Keywords: basic science, mega projects, Human Genome project, retrospection, epigenetics, exobiology, comparative evolution.

Introduction

Science projects come in all shapes and sizes, however an unusual and complex species of project has been evolving since the World War II. The new beast is huge, complex, multidisciplinary, and very expensive with regard to its care and feeding. "Big Science" megaprojects since the Manhattan project have tended to be multi-national, multi-centered, and multi-*billion* dollar affairs. In part because of these extreme costs, governments and agencies have become progressively more interested in collaborating to distribute costs and risk. Another motivation for participants is to leverage access and to establish a stake in the future benefits that will accrue. Scientists have consequently been careful, in the pitch to sell a project to funding agencies, to emphasize the promise of future technical advantage and the importance that these projects hold for the training and development of new scientists, engineers, and industries. Improvement of the technical and intellectual economy is usually highlighted.

Although there can be substantial overlap, "Big Science" projects come in two broad varieties: applied and basic. In applied science, recent knowledge is marshaled to assemble new application-oriented *projects* that can be exploited for profit in the immediate future: "how can I use or implement this now?" By contrast, although basic science also relies upon prior knowledge, the work encompasses a historically broader span, and the enterprise develops hypotheses that will lead to *programmatic* advancement over a more lengthy future.[1]

In the category with a basic science emphasis, "Big Science" projects occupy a dichotomous space. The huge costs are justified using the enlargement of potential for future hypothesis building, training, and applied science spin-offs. However, despite political representations that emphasize the futuristic and long-lived utility of the knowledge being sought, these programs are remarkable for often focusing upon ancient, even originary states. To some extent, we should expect this: even future-oriented projects will deliver insights into our present state and how the past must have unfolded to bring us here. Indeed it makes teleological sense that scientific progress will be facilitated by constant refinement of our understanding of both present and past states—the predictive value of a hypothesis is increased when we texture it to account for as many impinging factors as possible. Nevertheless, I will argue that even the projects with an ostensibly future-oriented prospectus either already possess the seeds of a *profound* retrospective emphasis from their inception, or they soon develop it. Even those projects that are described almost

[1] An example of Basic Science can be found in Roentgen's meticulous 1895 studies of the invisible light emitted by a Crooke's tube: this light could pass through solid matter and expose sealed photographic plates. This spectrum of light, called "X-rays," was of intense scientific interest, and he demonstrated their properties by displaying a crude image of his wife's hand—bones and wedding ring exposed for the viewing. The Applied Science implementation started as soon as the following year when Cannon demonstrated the practical utility of a Crooke's tube by locating a bullet in a Rochester NY patient. Coolidge subsequently refined the technology in 1903 by providing a tungsten filament and cooled anode as a target for the generation of X-rays.

exclusively in terms of future potential inevitably turn to the long distant past, and make it a major focus of the enterprise.

"Big Science" asks big and fundamental questions: the initial phase of cataloging is vast and is followed by a period of hypothesis development and testing that asks "how does this/we work?" Not surprisingly, the culmination becomes the question "how did this/we originate?" For example, the Hubble and Planck telescopes look into the far reaches of space and, in the process, they literally look back though time (Table 1). The Planck microwave map will indicate the state of the universe around 360,000 years after the Big Bang—some 13.3 to 13.9 billion years ago. Similarly, in an effort to find the Higgs boson, the $6 billion Large Hadron Collider was constructed to recreate conditions thought to have existed very shortly after the Big Bang. One of the few exceptions to the retrospective mega-projects, at least with regard to how it was sold to the public, politicians, funding agencies, and indeed to the scientists themselves, is the Human Genome Project.

Our first order of business here will be to show how the Human Genome Project was initially sold as a future-oriented informational *cornucopia* (or is that pharmacopeia?), and how it has now reflected back upon itself to emphasize comparative evolutionary studies and originary states. This late-found retrospection includes questions such as "what is the evolutionary origin of genes and of multi-cellular organisms such as ourselves?"

Second, we will see that even though subsequent and developing spin-off projects *appear* to have a future-oriented sales pitch, they too inevitably focus upon ancestral or originary states. In biology, the Post-Genome era is moving towards comprehensive studies of genome products (the Transcriptome and the Proteome), and how each interacts with the other and with the environment (the Epigenome). Our example will be the Human Epigenome Project which was initially defined as a future-oriented program, but that already has within it the seeds for a retrospective compulsion. As the likely candidate for a new "Big Science" project, I will outline how the Epigenome Project's first few reductionist and future-oriented steps of "how-does-it-work-science' coalesced to a comprehensive cataloging effort that inevitably focused attention upon the far distant past.

Finally, we will turn to a likely candidate for a future mega-science project, Exobiology (the study of life on other planets) and show how it has become profoundly concerned with origins even before it has got off the ground, so to speak, and launched into the future of space exploration. It seems that even futuristic sciences like Exobiology are looking retrospectively to evolutionary origins for inspiration.

Table 1. List of "Big Science" Project sponsors, planning to completion dates, costs, orientation and temporal focus.

Project/Program	Plan – Finish Date	Agencies	Estimated Cost (billions)	Applied/Basic Emphasis	Articulated Temporal Orientation
Manhattan	1939–1946	U.S., U.K., Canada	$1.8	Applied	Future
Moon/Apollo	1961–1972	U.S.	$170	Applied	Future
Shuttle	1968–2011	U.S. (NASA, DoD) with minor participation from CSA, ESA, Germany	$174	Applied	Future
Space Station	1992–2020?	International (NASA, ESA, CSA, RKA, JAXA)	$35 to $160	Applied	Future
ITER Fusion Reactor	1985–2036	International (U.S., E.U. Japan, S. Korea, Russia, India, China	$18–$20	Applied	Future
Large Hadron Collider	1977–present	E.U.	$9	Basic	Past
Hubble Space— Telescope	1990–2013?	U.S. (NASA) E.U. (ESA)	$6	Basic	Past
Planck/Herschel Space Telescope	1994–2012	U.S. (NASA) E.U. (ESA)	$2.5	Basic	Past
Webb Telescope	2004–2019	U.S. (NASA), Canada (CSA), E.U. (ESA)	$4.5	Basic	Past
Human Genome	1986–2006	International (9 countries formally)	$2–$3	Basic	Future?
Epigenome	2004–	U.K., France, Commercial	$0.13	Basic	Future?
Exobiology		?	?	Basic	Future?

The major projects of the first half of the last century have tended to be applied in nature and to have been propelled by military and political considerations, or with a view to providing a platform for future basic science explorations. By contrast, international consortia with a basic science emphasis have been dominating the agenda over the recent decades.

The Human Genome Project—A Brief History of the Sales Pitch and Delivery

Despite the monumental scale of the vision and the monolithic size of the final project, at the outset, the Human Genome Project did not enjoy universal approval even among scientists. At a 1986 symposium at Cold Spring Harbor, where some of the titans of molecular genetics assembled to consider a systematic and co-ordinated approach to the genome, concern was expressed that the $3 billion price tag would be extracted at the cost of existing funding programs, science projects, and careers (Cook-Deegan 1994, 111). Moreover, doubts were expressed regarding the value of the revealed DNA sequence: "sequencing the genome would be about as useful as translating the complete works of Shakespeare into cuniform" (Walsh and Marks 1986, 590). In other words, we would have a series of letters without knowing how the syntax is assembled, how instructions are conveyed, or ultimately, how they translate into the way we are actually constructed. I was working in a developmental biology lab at the time, and although we were excited about how useful the information could prove, we railed that the genome sequence alone could never tell us how organisms are assembled.

Nevertheless, momentum and opportunity emerged, and the hard selling began. A prominent scientist proclaimed in an editorial of the preeminent journal *Science* that the Human Genome represented the most sensible way to characterize and understand cancer (Dulbecco 1986). Much of the language that was addressed to the lay community, though, tended to revolve around two themes: the grandiose, even God-like accomplishment that the enterprise would represent, and the huge practical and commercial future utility that the information would deliver.

For example, James Watson, co-discoverer of the structure of DNA and the first director of the Human Genome Project exclaimed "the genetic messages encoded within our DNA molecules will provide the ultimate answers to the chemical underpinnings of human existence" (1990, 44). "We'll have the ultimate description of life" (Gorner and Kotulak 1990, 114). When a first draft of the genome arrived, ahead of schedule and under budget, the BBC World News service declared "the date of 26 June, 2000, will go down as one of the great moments in human history" and the article then went on to deliver the words of enthused scientists including John Sulston, a pioneer in the field: "...we are going to hold in our hands the set of instructions to make a human being" (2000). The second director of the Project, Francis Collins, echoed these sentiments: "It is humbling for me and awe inspiring to realize that we have caught the first glimpse of our own instruction book, previously known only to God" (2000). Evidently, the pitch resonated: President Bill Clinton responded with the words "today we are learning the language in which God created life" (2000).

Given the projected expense, it is not surprising that the Project's long-term medical and commercial potential was highlighted. In the journal *Science*, Collins and colleagues outlined the argument: the enterprise was the "single most important project in biology and the biomedical sciences" and would revolutionize "the breadth and depth of biological questions that can be addressed." Moreover, it would have "unprecedented impact and long-lasting value for basic biology, biomedical research, biotechnology, and health care" and for the "future prevention and treatment of illness" (Collins et al. 1998, 682). For Watson (2000), the project embodied "a giant resource that will change mankind, like the printing press." The information bonanza was envisaged to transform molecular medicine (cloning, gene therapy), pharmacogenomics, our understanding and treatment of cancer, and the development of exquisitely patient-specific diagnostic and treatment strategies. Sometime, in the near future, patients or their tumours could be genetically scanned, the disorder identified with absolute precision, and a tailor-fit treatment designed. Again, politicians expressed approval with thematic resonance. In 2000, President Clinton declared that the Genome Project would "revolutionize the diagnosis, prevention, and treatment of most, if not all, human diseases" (qtd. in Butler 2010, 1000), and Lord Sainsbury, UK Minister for Science amplified this with the words: "we now have the possibility of achieving all we ever hoped for from medicine" (2000).

The Seeds of Retrospection in the Human Genome Project

The speedy delivery of DNA sequence data has not been matched by an equally rapid translation to a pharmacopia of delights—quite the contrary. Indeed although a 2010 poll of scientists indicated that 69% were inspired to become a scientist or to change direction, and 90% felt that their own research had benefited from the project, there was nevertheless a general sense that potential benefits had been overhyped (Butler 2010, 1001).

The seeds of retrospection were present from the outset of the project. Perhaps in part to assuage potential competitors during the early days of consortium building, and in part because the Human Genome Project relied upon techniques developed for the study of other organisms, the final project came to encompass quite a diversity of organisms. Inevitably, genomes were compared, and inferences were made with regard to evolution and ancestry. In addition, as a consequence of huge technological advances, the cost of sequencing a genome as large as ours plummeted from $2 billion to about $1000.[2] Why not turn the technology to other organisms? Finally, as the true measure of genome complexity became apparent, the necessity of a comparative approach became paramount—in essence, a major question related to how we could parse the new genetic data if we didn't understand its syntax? Perhaps a comparison of diverse genomes could help us to elucidate general rules and to identify operational units. Comparative genomics therefore became a major interest, and this in turn provided fodder for studies already under way concerning operational rules, evolution, and ancestry. The genomics enterprise was turning from future potential to evolutionary past.

Along the way there were substantial surprises. For example, the human genome constitutes far fewer genes than originally thought—now estimated to be in the neighborhood of 24,000 (Lander et al. 2001), roughly a thousand genes less than the mustard-related plant *Arabadopsis* (Arabidopsis Genome Initiative 2000). Moreover, sponges possess almost all of the species and families of genes encoded within the human genome, and in the sponge these number about 18,000 (Srivastava et al. 2010). Even the lowly (and anatomically very simple) nematode worm has 19,000 genes (The *C. elegans* Sequencing Consortium 1998). How then, do we account for the huge differences in complexity amongst these organisms? Why aren't we more like sponges and nematode worms? For that matter, why do we retain the baggage of prehistoric retroviral invasions: nearly 42% of our genome represents the remnants of ancestral infections (Lander et al. 2001)? Surely evolution would have exerted pressure to reduce the wasteful reproduction of "junk" DNA. Just as T.S. Eliot relates how exploration into the unknown leads us back to where we started so that we can know the place for the first time, the Human Genome surprises revealed just how little we knew and forced us to substantially revise our understanding of its structure, activity, and evolution. Not just once . . . but several times . . . This recursive and retroactive revision tended to reinforce

---

2 This is the estimated cost for sequencing human genome in its entirety; however the process relies upon previously established maps provided by the Genome Project as a scaffold for alignment and assembly.

comparative evolutionary approaches, and in the absence of equally impressive leaps forward in the medical and pharmacological arenas, the evolution of genomes has assumed programmatic prominence.

The imaginative forays recently made into evolutionary genomics are numerous. Belinda Chang and colleagues at the University of Toronto compared the sequences for a visual pigment necessary to eye function in diverse limbed animals. She then worked backwards to infer the sequence of the gene belonging to their last common ancestor. This was an *Archosaur* that roamed the Earth 240 million years ago during the early Triassic Period. She then synthesized the adduced gene in the lab, used it to produce the prototype's visual pigment, and finally she and her colleagues tested that pigment for light sensitivity. Apparently *Archosaurs* could see further into the infrared than we are able—you wouldn't want to slink around a hungry one at night (Chang et al. 2002)!

Nor have evolutionary comparisons focused entirely upon the biologically remote past: more proximal comparisons have been made between humans and other primates. These have provided fodder for interesting speculation and experimentation. For example, perhaps we owe our ability to speak (and the rapid evolution of our societal and intellectual complexity) to a few mutations in select genes like *FOXP2* (Enard et al. 2002). Additionally, a preliminary draft of the Neanderthal genome has been assembled, and evolutionary inferences made (at some point we interbred!) (Green et al. 2010). Why stop there? The genome of the wooly mammoth has been resurrected (Miller et al. 2008), and there is talk of cloning and rearing a living one. Surely Jurassic Park is only a step around the corner! As an example of ostensibly future-oriented basic science megaprojects, the Human Genome and ancillary sequencing projects seem to have become inordinately and obsessively retrospective.

A Big Science Project in the Building Phase—The Nascent Human Epigenome Consortium

Recently, a new project has been consolidating. In some ways, it can be viewed as an offshoot of the Human Genome Project, but instead of focusing upon genome sequence, the objective is to define the modifications that are made to it as it is packaged. DNA packaging determines the freedom with which it can interact with proteins and other molecules—it regulates whether genes are accessible and can be turned on or off. Differences in packaging are what give our chromosomes a banded appearance, and frequently, entire clusters of genes are regulated as a cohort by alterations to packaging within a chromosomal segment. This type of regulatory change is termed epigenetic. It affects gene behaviour directly, by subtly modifying the chemistry but not the DNA sequence, as well as indirectly, by changing the proteins and RNA molecules with which DNA associates. These packaging changes are thought to begin with the addition of tiny methyl groups to one of the four nucleotides that comprise DNA, namely the "C" or cytosine. This "decoration" by methylation in most cases seems to nucleate the recruitment of local factors to make more profound changes to the proteins around which DNA is coiled, especially the histones. The end result is that the conformation of DNA is altered, the assembly of DNA with proteins (chromatin) is modified, and the associations that a gene has with the histones become restrictively intimate. Genes with many C's methylated tend to be inactive. By contrast, those genes that are under-methylated tend to be active—they can serve as the template to make RNA which can ultimately lead to the synthesis of protein. These different epigenetic states are also termed an epigenetic "imprint."

Epigenetic imprints, like DNA sequences, are heritable: not only do you inherit genes from a parent, but you also inherit patterns of how those genes are packaged, and therefore patterns regarding their behaviour.[3] Imprints can differ depending upon whether they are inherited from a mother or a father. Incredibly, life experience itself can modulate an imprint, resulting in heritable changes to gene activity. Events early in life can lead to gene behaviours later with implications not only for your own future health, but also for that of your offspring. An example of this can be seen in the inherited propensity to low birth weights and predispositions to certain diseases in the offspring of parents or grandparents subjected to natural or war-induced famines (Pembrey 2002; Roseboom, de Rooij and Painter 2006; Heijmans et al. 2008). In a very practical sense, the epigenome—patterns of gene imprinting and behaviours—represent a link between the genome, an environment, offspring, and disease.

The objective of the Epigenome Consortium is to focus primarily upon the initial step of DNA methylation and to develop a genome-wide methylation

[3] While imprinting does not alter the sequence of DNA, its heritable nature does introduce a surprising element of Lamarckism to our modern conception of heritability. Imprinting is now an important element widely investigated in cross-generational disease studies and cancer biology.

map against which patient samples can be compared for prognostic purposes. The anticipation here, as for the Genome Project, is that specific disease states and proclivities can be identified and treatment regimens tailored to fit. Sound familiar? The same objective was articulated for the Human Genome Project.

Epigenome Research is Already Concerned with Origins

It would be a mistake to limit the temporal scope of this latest initiative in epigenome studies to the origin of a disease state or proclivity. As the broader field of epigenome science has evolved, it has demonstrated an inherently evolutionary concern, and not merely because the epigenetic imprint is heritable. Central questions have revolved around what purpose imprinting serves, and how it evolved. For example, you inherit two copies of most genes, one from each parent. A subset of these genes comes with a different imprint, depending upon the parent of origin. Two genes, Insulin-Like Growth Factor 2 (IGF2) and its receptor (IGF2r), are differentially imprinted according to the parent from whom you inherit them. IGF2 encourages early fetal growth. The receptor plays a controlling role, and if it binds to IGF2, it presents the growth factor to cellular machinery for degradation. In short, they play antagonistic roles in the regulation of growth (summarized in Table 2).

Moore and Haig interpreted this to indicate that the parental genomes are in competition (1991). While both parents have an interest in healthy offspring, they differ subtly. Males have an interest in big and healthy offspring that are robust enough to survive and that might, perhaps, require sufficient energy from a mother as to divert her from other interested males. Mothers too have an interest in delivering healthy offspring, but not of a size to make inordinate demands upon their metabolism and health. No huge heads to deliver through the birth canal—it would be nice to survive to reproduce again thank-you very much!

There are alternatives to this neo-Darwinist interpretation, however. Arguments have been advanced to suggest that imprinting plays a role in the evolution of mate fitness, as well as to inhibit the lethal effect that an ectopic pregnancy would have if accidentally stimulated eggs were to grow in the absence of sperm (Morison, Ramsay and Spencer 2005). Additionally, imprinting might have evolved to ensure minimal metabolic investment in embryos that are missing chromosomes: these would fail early and miscarry before they absorb much in the way of maternal invest-

Table 2. The relationship between the genes' activities, their imprint status, and role in early fetal growth depends upon from which parent the gene is inherited.

Gene	Parent of Origin	Methylation status	Gene Activity	End Result
IGF2	Father	low	high	growth
	Mother	high	low	no role
IGF2r	Father	high	low	no role
	Mother	low	high	constrained

(Adapted from Moore and Haig 1991.)

ment (Wilkins 2005). A final possibility is that early in the evolution of the genome, invasion by retroviruses was so disruptive to chromosome stability that a mechanism to clamp down upon the invading sequences evolved. Interestingly, the long terminal repeat sequences of these viral insertions are very rich in cytosine or "C" specifically in combination with G's, and this is the where C's tend to be targeted for methylation, thereby rendering the sequence more susceptible to inactivation (2005).

An interesting development in recent years has been that a plant protein involved in carbohydrate metabolism, DEMETER, establishes the imprinting ability of another gene that is itself imprinted, namely MEDEA (Gehring and Henikoff 2007). Carbohydrate metabolism is immensely varied and complex, and is suggested to have played a central role in life's origins (Stern and Jedrzejas 2008). There is no aspect of cellular and organismal life that is untouched by carbohydrate metabolism, from the trafficking of proteins, nutrients, and wastes, to the activation/inhibition of critical cell cycle and signaling players in a cell, or to the fecundity of seeds... even to lesbianism in mice. The linkage of the two processes, carbohydrates and imprinting has interesting implications for the unfolding discussion. Clearly, just as the Genome Project expanded our appreciation of evolutionary possibilities and processes, characteristics of the epigenome are providing fodder for retrospection to origins as well.

Exobiology—Will the Next Big Science Project Be Retrospective?

Intuitively, nothing seems more futuristic than Exobiology—the study of life on other planets. As a present- and future-oriented enterprise, this science has also already become profoundly retrospective. Our tentative forays into space to look for life have been limited by our capacity to make direct contact with other celestial bodies. Venus, Mars, and the Moon define the range of our current repertoire. Launching to and directly investigating other planets remain largely in the realm of science fiction: we are currently restricted to making indirect inferences from a distance. A big question in everybody's mind has to be "if we see alien life, how will we recognize it?"

Paul Davies and colleagues advocate starting the search much closer to home, but in a sense, further back in time. They argue that other forms of life might already have been launched on Earth, and that the place to go looking for "life as we do not know it" is in places with extreme and isolated environments (Davies et al. 2009). Examples of such places are marine thermal vents, deserts, deep in the Earth's crust, or in poisonous terrestrial and aquatic ecosystems. A controversial NASA-affiliated study illustrates the possibilities for success using this approach: extremophile bacteria have been described recently that appear to substitute arsenic for phosphorus in their DNA (Wolfe-Simon et al. 2010). This highly contentious observation has proven difficult to repeat; however if true, it suggests that the molecules that encode and regulate life can, and *already have* taken very different forms on Earth. In short, the agenda for developing the tools to search for new life in the future is to search in the very places where we ourselves are thought to have originated billions of years ago.

Evolution and Exobiology—Where Future, Past, and Present Meet

One possibility not yet addressed, and that I will have the cheeky presumption to advance, is that one place to look for remnants of alien life might be within our very selves. There is precedent for such a search. Open any basic biology textbook and you will find reference to the origins of mitochondria and chloroplasts, the "power houses" of animal and plant life respectively, as endo-parasites. The prevailing theory is that their physical and metabolic presence was co-opted over time and evolution to the construction of a symbiotic and eventually unified whole. Both mitochondria and chloroplast retain remnants of their former genomes, and our chromosomally-centered genomes pick up the rest. One wonders if epigenetic imprinting is a remnant of a formerly antagonistic relationship? Perhaps imprinting reveals a competition that is much older than the one between maternal and paternal interests, and perhaps parental imprinting effects are, like mitochondria and chloroplast, co-opted indicators of former parasites?

For that matter, might there originally have been two radically different forms of life on the planet, one nucleic acid based, and one carbohydrate based? Given some of the counter-regulating mechanisms within existing living systems, this might be a pertinent question to ask. From the interaction of DEMETER and MEDEA in plants discussed earlier, we know that epigenetic packing is, at least indirectly, affected by carbohydrate metabolism. Additionally, after genes are activated and their encoded proteins are synthesized, these latter are often regulated by a process called phosphorylation. The very sites that are attractive for phosphorylation are also often bound by modified carbohydrates called O-linked sugars, and indeed, the two may operate competitively (Chou, Hart and Dang 1995). Moreover, phosphorylation is often one of the first steps in the degradation of sugars. Finally, it is tempting to speculate that the shuttling of carbohydrates to the external cell surface (requisite to the deposition of extra-cellular matrix in skin, bone, and virtually any other tissue), as well as the formation of a thick cell wall in plants and bacteria, represents the evolutionary consequence of a co-opted battle for dominance. The carbohydrates were, at least partially, locked out!

Conclusion

The observation that I have made is, in essence, a very simple one: the mega-projects of basic science are profoundly retrospective no matter what the espoused agenda and temporal orientation. A degree of retrospection should not have been, from the outset, all that surprising. After all, when we address fundamentals, it is inevitable that discourse turns to origins—complexity has to start somewhere. I have argued that the pressure of recursive teleological revision, especially when big surprises are revealed by new technologies, enhances the re-direction of interests from future to past. Moreover, when the prospects for innovation and reward in the future begin to appear less immediate, other interests and opportunities occupy the vacuum.

If there is a theme to be derived from the commonality of retrospective interests that develop in "Big Science," it is that the frontier of human knowledge still appears to be driven by a need to probe introspectively: "who are we, what are we, where did we come from, and why are we here?" Not incidentally, the compulsion to pursue these questions reveals another unexpected, but comforting recognition. All of the "Big Science" projects assembled over the past few decades have been progressively more international, interdisciplinary, and collaborative in scope: they also forefront a quest for *peaceful* knowledge.

References Cited

Arabidopsis Genome Initiative. "Analysis of the Genome Sequence of the Flowering Plant Arabidopsis Thaliana." *Nature* 408, no. 6814 (2000): 796–815.

Butler, D. "Human Genome at Ten: Science after the Sequence." *Nature* 465, no. 7301 (2010): 1000–1001.

Chang, B.S., K. Jonsson, M.A. Kazmi, M.J. Donoghue, and T.P. Sakmar. "Recreating a Functional Ancestral Archosaur Visual Pigment." *Mol Biol Evol* 19, no. 9 (2002): 1483–1489.

Chou T.Y., G.W. Hart, C.V. Dang. "c-Myc is glycosylated at threonine 58, a known phosphorylation site and a mutational hot spot in lymphomas." *J. Biol. Chem.* 270, no. 32 (1995): 18961–18965.

Clinton, President William, "What They Said: Genome in Quotes." *BBC World Service* (26 June 2000). http://news.bbc.co.uk/2/hi/science/nature/807126.stm (accessed 25 September 2010).

Collins, Francis, "What They Said: Genome in Quotes." *BBC World Service* (26 June 2000). http://news.bbc.co.uk/2/hi/science/nature/807126.stm (accessed 25 September 2010).

Collins, F.S., A. Patrinos, E. Jordan, A. Chakravarti, R. Gesteland, and L. Walters. "New Goals for the U.S. Human Genome Project: 1998–2003." *Science* 282, no. 5389 (1998): 682–689.

Cook-Deegan, R.M. *The gene wars: science, politics, and the human genome*. New York: W.W. Norton & Co, 1994.

Davies, P.C., S.A. Benner, C.E. Cleland, C.H. Lineweaver, C.P. McKay, and F. Wolfe-Simon. "Signatures of a Shadow Biosphere." *Astrobiology* 9, no. 2 (2009): 241–249.

Dulbecco, R. "A turning point in cancer research: sequencing the human genome." *Science,* 231, no. 4742 (1986): 1055–1056.

Enard W., M. Przeworski, S.E. Fisher, C.S. Lai, V. Wiebe, T. Kitano, A.P. Monaco, S. Pääbo. "Molecular evolution of *FOXP2*, a gene involved in speech and language." *Nature* 418, no. 6900 (2002): 369–872.

Gehring, M., and S. Henikoff. "DNA Methylation Dynamics in Plant Genomes." *Biochim Biophys Acta* 1769, no. 5–6 (2007): 276–286.

Green, R.E. et al. "A Draft Sequence of the Neandertal Genome." *Science* 328, no. 5979 (2010): 710–722.

Gorner, P. and R. Kotulak. "Biology Goes For It All By Mapping The Human Genetic Code." *Chicago Tribune* (8 April 1990): 114.

Heijmans, B.T., E.W. Tobi, A.D. Stein, H. Putter, G.J. Blauw, E.S. Susser, P.E. Slagboom, and L.H. Lumey. "Persistent Epigenetic Differences Associated with Prenatal Exposure to Famine in Humans." *Proc Natl Acad Sci USA* 105, no. 44 (2008): 17046–170469.

Lander, E.S. et al., "Initial Sequencing and Analysis of the Human Genome." *Nature* 409, no. 6822 (2001): 860–921.

Miller, W. et al. "Sequencing the nuclear genome of the extinct woolly mammoth." *Nature* 456, no. 7220 (2008): 387–390.

Moore, T., D. Haig. "Genomic Imprinting in Mammalian Development: A Parental Tug-of-War." *Trends Genet* 7, no. 2 (1991): 45–49.

Morison, I.M., J.P. Ramsay and H.G. Spencer. "A Census of Mammalian Imprinting." *Trends Genet* 21, no. 8 (2005): 457–465.

Pembrey, M.E. "Time to Take Epigenetic Inheritance Seriously." *Eur J Hum Genet* 10, no. 11 (2002): 669–671.

Roseboom, T., S. de Rooij, and R. Painter. "The Dutch Famine and Its Long-Term Consequences for Adult Health." *Early Hum Dev* 82, no. 8 (2006): 485–491.

Sainsbury, David Lord. "What They Said: Genome in Quotes." *BBC World Service* (2000; 26 June) (accessed 25 September 2010).

Srivastava, M. et al. "The *Amphimedon Queenslandica* Genome and the Evolution of Animal Complexity." *Nature* 466, no. 7307 (2010): 720–726.

Stern, R. and M.J. Jedrzejas. "Carbohydrate Polymers at the Center of Life's Origins: The Importance of Molecular Processivity." *Chem Rev* 108, no. 12 (2008): 5061–5085.

Sulston, John. "What They Said: Genome in Quotes." *BBC World Service* (26 June 2000). http://news.bbc.co.uk/2/hi/science/nature/807126.stm (accessed 25 September 2010).

The *C. elegans* Sequencing Consortium. "Genome Sequence of the Nematode *C. elegans*: A Platform for Investigating Biology." *Science* 282, no. 5396 (1998): 2012–2018.

Walsh, J.B. and J. Marks. "Sequencing the human genome." *Nature* 322, no. 6080 (1986): 590.

Watson, James. "What They Said: Genome in Quotes." *BBC World Service* (2000; 26 June). http://news.bbc.co.uk/2/hi/science/nature/807126.stm (accessed 25 September 2010).

Watson, J.D. "The human genome project: past, present, and future." *Science* 248, no. 4951 (1990): 44–49.

Wilkins, J.F. "Genomic Imprinting and Methylation: Epigenetic Canalization and Conflict." *Trends Genet* 21, no. 6 (2005): 356–365.

Wolfe-Simon, F. et al. "A Bacterium That Can Grow by Using Arsenic Instead of Phosphorus." *Science* (Dec. 2010). http://www.ncbi.nlm.nih.gov/entrez/query.fcgi?cmd=Retrieve&db=PubMed&dopt=Citation&list_uids=21127214 (accessed 25 September 2010).

THE PAST-FUTURE ASYMMETRY*

Friedel Weinert

Abstract: Since the past-future asymmetry—the fact that we have records of the past but not the future—is still a puzzle, the aim of this paper is twofold: a) to explain the asymmetry and its status in philosophy and physics and to critically review the proposed solutions to this puzzle; b) to advance a dynamic solution to the puzzle (which is lacking in alternative proposals) in terms of the 'universality' of the entropy relation in statistical mechanics.

Keywords: Asymmetry, block universe, decoherence, entropy, order/disorder, phase space, spreading, theory of relativity, thermodynamics, thermodynamic probability, time travel.

I. *Introduction*

"[E]nergy is Nature's currency..."
W. Alvarez, *T.Rex and the Crater of Doom* (1997, 8)

It is a commonplace that we can influence the future but not the past; that we have knowledge of the past but not of the future; and that in most accounts time travel into the past engenders paradoxes, which do not arise from time travel into the future. There exists therefore a *past-future asymmetry*. Although it is a commonplace, the asymmetry is, on reflection, also puzzling, since no such problems exist in the spatial analogue. We can return to a place we have visited in the past an indefinite number of times; we can alter a past place, and an architect's plan incorporates knowledge of future places. Yet, in the temporal analogue, we are powerless with respect to the past, if not with respect to the future. There is no here-there space-asymmetry in the same way as there is a past-future time asymmetry. But why can we not visit past times in the same way that we can visit past places? A simple answer, inspired by presentism, is that the past no longer exists and the future has not yet arrived. Only the present is real. But this simply postpones the puzzle: why does the past no longer

* The author would like to thank the editors of this volume and two anonymous reviewers for their constructive and helpful comments on an earlier version of this paper.

exist? And when does the present exist? Saint Augustine may have a point when he says that the present moment is infinitely small for any event that is Now—say, the utterance of a word—can be subdivided into past and future moments. Note that the simple answer does not hold for the spatial analogue. For it could reasonably be said that the past exists in fossil records and old monuments. I can lay my hands on the Sphinx in a way that I cannot lay my hands on Bismarck. It should also be noted that the simple answer presupposes a dynamic view of time. It contradicts, as we shall see, the Special Theory of Relativity, which has destroyed the notion of a universal Now and seems to imply the so-called block universe, in which past, present and future are equally real. The block universe implies a static view of time. So does the widespread view that the fundamental laws of physics make no distinction between past and future, that they are time-reversal invariant.

Philosophers have provided a number of answers to the puzzle of the past-future asymmetry. The purpose of this paper is to review some of these standard answers and to revisit the question by appeal to the 'universality' of the entropy relation in thermodynamics and statistical mechanics. The discussion will necessarily be limited to the familiar macro-world because it is in this world that humans experience the past-future asymmetry. If this experience can be explained by appeal to the notion of entropy, it is this notion as it appears in standard textbooks of statistical mechanics that must carry the explanatory weight. (Connections of entropy with quantum mechanics and cosmology will be briefly discussed in the footnotes.)

II. *The Block Universe*

There is a prominent view amongst physicists and philosophers—the block universe—in which the past-future asymmetry does not arise. This view harks back to antiquity (Parmenides) but has gained much support from the Special Theory of Relativity. As the Special Theory of Relativity is not a cosmological theory, it does not consider the expansion of the universe and the possibility of a cosmological arrow.[1] The block universe is

[1] It should be kept in mind that according to the latest cosmological discoveries, the universe not only expands but expands at an accelerated pace (cf. Schmidt [2005]. Already in 1929 Edwin Hubble discovered a linear relationship between redshift and the distance

inferred from the results of the Special Theory. According to this view, the physical world simply exists and the passage of time is a human illusion or at least a product of human perception (cf. Norton 2010). Imagine that the physical world is totally determined in the sense that today's state of the world determines tomorrow's state, just as yesterday's state of the world determined today's state, *ad infinitum*. Human beings are too limited to survey the physical state of the universe at its various stages, so they are forced to view the world stages sequentially, which gives the impression of the flow of time. But this passage of time is merely a result of our perceptual limitations. The famous demon invoked by Pierre Simon Laplace (1820) is able to see the vista of world history in one fell swoop. For the demon of superhuman intelligence, all the stages of world history are already recorded like frames on a film. It is only the weakness of human perception and intellect that requires humans to see the frames sequentially. If the laws of physics determine the states of the universe, from A to Z, then these states already exist, in the past and in the future, irrespective of whether they have occurred (or not) or have been observed. By way of an analogy, consider that a mighty demon has singled you out to win next Saturday's jackpot in the national lottery. Then you already are the winner in the eyes of the demon, although for you the event has not yet occurred. In this scenario then the demon possesses equal knowledge of past and future affairs; human beings come closest to such knowledge in the laws of physics. Still, this demon would not be able to change the past, but time travel into the past and the future seems to constitute no problem.

Immanuel Kant arrived at a similar view. In the Kantian view, time is a pure form of intuition—a condition of the mind that apparently imposes a temporal order on appearances. The result of this conception is that in the absence of humans, there is no time, no temporal order of things in themselves:

> It [time] is no longer objective, if we abstract from the sensibility of our intuition, that is, from that mode of representation, which is peculiar to us, and speak of *things in general*. (*Critique* 1787, B51/A35)

Kant is worth mentioning in this connection because many physicists thought that the Special Theory of Relativity endorsed a Kantian view of time. Under this scenario of the block universe, let us consider time

---

of galaxies, which is known as Hubble's law: $v=Hr$, where v is the radial velocity, r is the distance, and H is Hubble's constant.

travel into the future and time travel into the past; such a discussion will throw light on the past-future distinction. Time travel is only possible if the block universe exists, and its consideration will equip us with tools to enhance our understanding of the past-future asymmetry.

III. *Special Theory of Relativity*

III.a. *Time Travel into the Future*

Time travel into the future is a popular fiction, the best example of which is H.G. Wells's *The Time Machine*, in which a time traveller leaves his own time (1895) to travel into the distant future and then return to his own present. Time travel into the future does not involve the kind of paradoxes that beset time travel into the past. But it involves certain oddities because time travellers into the future can learn of future discoveries, which they can bring back to their own time. This possibility is illustrated by the tale of a time-travelling mathematician (Davies 2001, 114). This mathematician struggles for years without success to prove an important mathematical theorem. One day he decides to build a time machine to explore if future generations have found a proof. He strikes it lucky and finds an elegant proof in a mathematical journal published in the future. He returns to his own time and publishes the proof. The assumption, admitted in the block universe, is that the future already exists. Its discoveries and inventions can be harvested by a time traveller into the future. Time travelling into a pre-existing future is also a central feature of Wells's famous book, *The Time Machine* (1895). Wells operated with a notion of space-time that is reminiscent of Minkowski space-time and central to the Special Theory of Relativity. The latter involves the so-called twin paradox, which is sometimes presented as a form of time travelling into the future.

But there are important differences between the Wellsian and the relativistic notion of space-time.

The Wellsian time traveller travels along the temporal axis that exists as a fourth dimension in a spatial sense. Wells treated time as a fourth dimension in addition to the three spatial dimensions:

> There are really four dimensions, three of which we call the three planes of Space, and a fourth, Time. There is, however, a tendency to draw an unreal distinction between the former three dimensions and the latter, because it happens that our consciousness moves intermittently in one direction along the latter from the beginning to the end of our lives. (2005, 4)

The Wellsian time traveller uses his knowledge of future history to return to his own present and to warn people of the detrimental evolution that has split the human race into the gentle Eloi and the ferocious Morlocks. The Wellsian time traveller uses time in a spatial sense, which means that it is a temporal road on which the time traveller can freely travel up and down, spanning centuries. The time traveller already hints at the existence of a Parmenidean block universe and a Kantian notion of time before these notions became associated with the Special Theory of Relativity. According to Wells, then, the only difference "between Time and any of the three dimensions of Space" is that "our consciousness moves along it" (2005, 4, 94). Of the various snapshots of a person's life, he says that they are "Three-Dimensional representations of his Four-Dimensioned being, which is a fixed and unalterable thing" (2005, 5).

On first approach the Wellsian time traveller seems to be engaged in a form of future time travel, which is similar to a well-known scenario by the name of the twin paradox in the Theory of Relativity. Two twins, Homer and Ulysses, decide to confirm the time dilation prediction of the Special Theory of Relativity. Homer will stay at home, while Ulysses will go on a space journey to a nearby star some eight light years away. Ulysses will travel at a velocity of v= 0.8c, apart from short bursts of acceleration and deceleration. According to Homer's clock, it will take twenty years for Ulysses to complete his journey—ten years for each leg—so that Homer will be twenty years older. But for Ulysses the journey will only take 12 years, according to his clock that measures time t' since his clocks slow down, as seen by Homer's clocks. If longer journey times were chosen for Ulysses to more distant planets and higher velocities, the temporal discrepancies between the earth inhabitants and the space traveller could amount to thousands of years (Lockwood 2005, 48; Novikov 1998, 77). Ulysses would then return to an earth where history had advanced for millennia. He would, as some conclude, have travelled into the future, whilst aging only a moderate amount of time (Davies 2001, 15–17). But Ulysses, on his return to earth, is as much a prisoner of time as the earth inhabitants he finds. He will not be able to return to the present, from which he started, as does the Wellsian time traveller; he will not be able to affect the course of history on earth after his voyage into space. "There are no contradictions involved in Ulysses' travel. Both Ulysses and Homer move forward in time but the motion is slower for Ulysses than it is for Homer" (Novikov 1998, 232). In other words it is not real time travel into the future in the Wellsian sense.

Wellsian space-time is quite different from the representation of time in the theory of relativity where space and time become welded into a union of space-time, and hence we speak of Minkowski space-time. In Minkowski space-time, spatial and temporal lengths become subject to relativistic effects, known as time dilation and length contraction. For the Wellsian time traveller, time is just another dimension in the geometry of space, which enables him to return to the year 1895, from which he departed far into the distant future. The time traveller experiences no relativistic effects. He enters his time machine in his own time and then observes as his particular location changes through the centuries, as his time machine accelerates along the temporal highway. But the theory of relativity has led to the same inference that Wells draws. The "four-dimensional being"—"fixed and unalterable"—is the block universe. But different observers 'slice' this block differently, which leads to disagreements about their respective clock times and the simultaneity of events. According to the Special Theory, there exists no universal Now. Thus an inference is drawn from the results of the Special Theory of Relativity to the atemporality of the physical world. The latter is associated with the block universe:

> Due to the absence of an observer-independent simultaneity relation, the Special Theory of Relativity does not support the view that 'the world evolves in time'. Time, in the sense of an all-pervading 'now' does not exist. The four-dimensional world simply is, it does not evolve. (Ehlers 1997, 198)

The four-dimensional world does not evolve, which means that its states stretch both into the future and into the past, as seen from the present frame. Some limited form of time travel into the future seems possible, given the time dilation effect, as illustrated in the twin paradox. But a Wellsian time traveller would encounter a number of oddities, like the exploitation of future knowledge, for his own gain in the present. The question then arises, given the block universe, whether the Special Theory permits time travel into the past and how it would deal with the paradoxes that arise.

III.b. *Time Travel into the Past*

Can Ulysses travel into the past? While time travel into the future does not seem to involve paradoxes, time travel into the past certainly abounds with them. To see the difference between mere oddities and paradoxes, consider Michael Dummett's story of a time traveller *from* the future who returns to the twentieth century to visit a painter who, in the time traveller's own era, is regarded as a great artist (1986, 155). When the time traveller sees the artist's work, he is disappointed and concludes that the artist is still to paint the paintings for which he will become famous in a future era. The time traveller shows the painter a book of reproductions of these famous paintings but then has to leave in a hurry, leaving the catalogue behind. The painter begins to copy these reproductions onto canvas. There are no obvious contradictions, yet the story strikes us as odd because it seems to involve no artistic creativity. The painter copies reproductions and produces paintings from reproductions. While the time-travelling mathematician finds a proof published in a future journal—it could have been produced by a future mathematician—there is no creativity involved in Dummett's tale. These are oddities that seem to leave open the conceptual possibility of time travel *from* the future. Note, however, that in both scenarios—time travel *into* and *from* the future—it is assumed that the future already exists, a permissible assumption in the block universe view.

Time travel into the past also comes in two flavours. One is cosmological time travel, which involves such fanciful scenarios as the splitting of universes, closed time-like curves, baby universes and wormholes (cf. Gott 2001; Carroll 2010). There are considerable technical problems associated with these scenarios but, at any rate, they do not amount to the free time travel presumably possible in conventional time machines. Conventional time travel, however, depends on the idea of time machines, which take Wells's time traveller into the past rather than the future. This kind of time travelling can lead to paradoxes. One, for instance, is the grandfather paradox. If time travelling into the past were possible, a time traveller could return to his own past, kill his relatives just before his birth, and thus prevent his own existence in the future. How then was he able to travel into the past? David Lewis maintains that time travel into the past is possible and that the paradoxes of time travel—like the grandfather paradox—are mere oddities. In the block universe view he can return to the past but he cannot change the past: "Time travel constitutes a possible pattern of events in a four-dimensional space-time, with no extra dimensions" (1976, 149. Also cf. Davies 2001, Ch. 4; Novikov 1998, 233–234). Time travel into the past is logically possible because it involves no genuine paradoxes. It is true, Lewis admits, that a time traveller cannot return to, say, the year 1921, kill his grandfather and change the past. But the word 'can' is equivocal. What the time traveller can do (in the past) is compossible with certain past facts but not with others. Thus, Lewis imagines that the time traveller can buy a gun, lie in wait for his grandfather, 'one winter day in

1921,' but he cannot kill his grandfather who died in fact in 1957. The time traveller cannot kill his grandfather because such an act is not compossible with the ensuing history of events. But, according to Lewis, a time traveller can visit the past as a passive observer who nevertheless performs certain compossible actions in the past, like observing his grandfather. Yet Lewis overlooks that many more facts about the time traveller fail to be compossible with the past than he notices. What his time traveller is permitted to do—buying a rifle in 1921, renting a room, kicking a stone, observing his grandfather—is already changing the past, since these actions are not compossible with certain physical facts about 1921.

The genuine time traveller is not part of the normal past. In the language of modern physics, macro-objects and their location "decohere" through the interaction with the photons in the environment (Schlosshauer 2008; Gell-Mann and Hartle, 1990, §5). All the macro-objects in 1921 decohered through the interaction with the environment. Hence a time traveller enters an environment that has already decohered all the objects of the past. A time traveller would therefore necessarily have to change the past because photons and background particles would start bouncing off her/him. He would therefore change the past through energy exchanges, but the past is fixed in *all* respects.[2] To appreciate the problem, consider the world of 1921 with the analogy of financial transactions, but instead of a monetary balance there is an energy balance. On the credit side there is a certain amount of energy available to do useful work; on the debit side there is a certain amount of lost energy (entropy) that can no longer be employed for useful work. Although the total amount of energy in the universe of 1921 remains the same, every time energy is 'moved' from the credit to the debit side the overall entropy balance of the universe changes. If Lewis's time traveller were to perform these acts in 1921, s/he would change the entropy balance of the world of 1921, of which the time traveller was not originally a part. Changing the entropy balance of the past means changing the past, but what is done cannot be undone. Thus the suggestion that "no paradox would ensue," as long as the time

[2] Atkins calculates that at room temperature (20°C) humans generate entropy at the rate of about 0.3 joules per Kelvin per second. One of the reviewers provided a helpful clarification: "One can calculate that a person who eats 2000kcal/day converts energy at the rate of 2000kcal/day × 4187J/kcal × 1 day (24 hrs × 3600 seconds) = 97 Watts. That is, a person converts energy at a rate comparable to a 100W light bulb. If this is all transferred ultimately to the environment's internal energy via heat processes, one gets the entropy generation rate 100J/300K = 0.33J/K/sec. to one significant digit, this confirms (and also clarifies) Atkins' estimate" (Atkins 2003, 118 FN 6).

traveller is "inactive and invisible" (Lucas 1973, 50) will not do because even an inactive time traveller, through the mere act of visiting, changes the entropy balance in the past. Even if the time traveller merely observed events in 1921, s/he would change the past because observation involves a photon interaction between a perceptual apparatus and an object. In the normal past, no time traveller's eyes spied on grandfather going about his daily routines. Photons bounced off objects in 1921 to be observed by grandfather's contemporaries. The past is fixed because there is no more decoherence left to do for the time traveller.

The time traveller could only change the past if he were part of it. A possible defence of time travel into the past is to say that the time traveller is already part of the past (Nahin 1993, 208). Lewis suggests that a "time traveller… is a streak through the manifold of space-time, a whole composed of stages located at various times and places"[3] (1976, 146). Such a suggestion is compatible with the view that objects in four-dimensional space-time exist as four-dimensional world tubes. Then the time traveller can indeed interact with people of the past (he is part of it), but he still cannot kill his grandfather. The suggestion involves the weird notion that a time traveller consists of "scattered stages" (Lewis 1976, 149). Yet it is not really time travelling in the sense of leaving one's own period to visit another period of time. Lewis claims that, from the point of view of the early stage of the time traveller's existence when he is part of the grandfather's present, the time traveller has the *ability* to carry out his murderous intentions since the future looks open to him and his act would be compossible with facts relevant to 1921, but some accident prevents him from doing so. Still from the point of view of the later stage of the time traveller, this act of killing is impossible because grandfather died in 1957. So the later stage must 'cause' the earlier stage to fail. Yet in the notion of the block universe, past, present and future are equally real and determined, so there is no need for backward causation. Either the time traveller has already scattered stages of existence in the past, in which case he is not a genuine time traveller, or the time traveller crosses the boundaries of his own period of existence, in which case he needs to interact with the past, and thus change it. In the latter case, Lewis's time traveller in a four-dimensional block universe neglects important thermodynamic features

3 Lewis's assumption of existence in stages implies that the time traveller's world line, which should start after 1921 with his birth, has earlier pieces in 1921, which must be cut off from his continuous world line stretching from his birth to his death.

about the physical universe that remain essential for a consideration of the past-future asymmetry.

Given these features, it would be preferable to consider the past-future asymmetry without reliance on the block universe because this view—like the common-sense view—comes with a package of philosophical commitments that skew the inquiry into the past-future asymmetry. The block universe is committed to the view that time is a human construct that derives from the well-known consequences of the Special Theory of Relativity: time dilation effects, the twin paradox, and the relativity of simultaneity. In this view, the physical world itself is timeless. But many commentators regard the Special Theory of Relativity as concerned only with kinematic features, at the expense of dynamic features that are part and parcel of the physical universe.[4] Hence a compelling need presents itself to understand the asymmetry between the frozen past and the open future without reference to the block universe.

[4] The claim that the Special Theory is only a kinematic theory that ignores dynamic features (Geroch 1978, 20; Janssen 2009; Born 1962, Ch. VI.7) must be made with some provisos in mind. It is customary, in textbooks on relativity, to treat the Special Theory as a kinematic theory only. But it is always possible to include dynamic aspects if questions are asked about the behaviour of clocks and rods in relativistically moving systems. Why do clocks appear to slow down, and why do rods appear to contract in systems that are moving inertly, at relativistic speeds, with respect to a system considered at rest? In order to answer such questions, dynamic considerations must be taken into account that refer to the behaviour of atomic systems that ultimately make up the macroscopic systems. The idea is that time dilation and length contraction may not simply be perspectival phenomena but may find an explanation on the quantum-mechanical level. Such a view was proposed by Wolfgang Pauli and lately defended by H. Brown (2005; see also the discussion in Janssen 2009). It must be stressed, however, that such an atomistic theory of length contraction and time dilation is not yet available. It is also common practice in physics to use idealizations and approximations, which means that a kinematic description is preferred at the expense of a dynamic description, which may either be unavailable or unhelpful for the discussion at hand. It is, therefore, a pragmatic question to what extent a relativistic system can be described kinematically only or also dynamically. In the standard interpretation, space-time is regarded as a fixed background and questions about length contraction and time dilation are merely perspectival features (Whitrow 1980, §5.5). But such a convenient separation of kinematics and dynamics no longer applies in the General Theory of Relativity because space-time itself becomes a dynamic entity. (The author would like to thank D. Giulini for discussion on these points.) The Special Theory can be dissociated from the static block universe in favour of a 'growing block' view, which has implications for the past-future asymmetry, but doing so would require reference, say, to thermodynamic features, as the author has discussed elsewhere (see Weinert 2010a; 2010b).

IV. *Some Attempts to Explain the Past-Future Asymmetry*

Are there any independent reasons to ground the past-future asymmetry? We will look at three different attempts to explain the past-future asymmetry that rely on more fundamental considerations and do not assume the 'truth' of the block universe.

1. Michael Lockwood (2006, Ch. 12) makes use of J.S. Mill's conditional model of causation, according to which the cause, *C*, of an event, *E*, can be broken down into a set of necessary and sufficient conditions. Jointly, the antecedent necessary and sufficient conditions can explain the occurrence of *E*. Lockwood introduces the term "nasic conditions," which is an acronym for Millian necessary and sufficient conditions in the circumstances. A causal set of antecedent factors, *C*, is a nasic condition for the occurrence of event *E*, where appropriate background conditions are taken into account. A cause introduces a change in the normal running of things. Lockwood furthermore embraces Reichenbach's observation that "one can infer the total cause from a partial effect but one cannot infer the total effect from a partial cause" (2006, 253; see Figure 1).

   Reichenbach himself points out that the "*inference from the partial effect to the total cause* is typical of all forms of recording processes"[5] (1956, 180; italics in original). Partial effects are typically sets of conditions that obtain locally in conjunction with other partial effects and general background conditions. Consider Karl Popper's pond analogy: when a stone is thrown into the centre of a pond it will create divergent waves that will break on the shore of the lake (1956, 538). From the arrival of divergent waves and the breaking of the waves on the shoreline, one can infer that a disturbance at the centre caused the divergent waves. The analysis of the divergent waves may tell us something about the physical properties of the object. But knowing only partial causal conditions prevents us from inferring what the total effect of these conditions will be. Thus, knowing that a stone has been launched at an angle α, and that some additional parameters are satisfied, we can calculate various characteristics of its trajectory; but if we do not have

[5] This conclusion does not necessarily hold under conditions of uncertainty about partial effects and when the cause-effect relationship is a many-to-one relation, as it may occur in irreversible thermodynamic processes. For instance, the present temperature of a gas near equilibrium may have multiple microscopic configurations as a cause. For a discussion see Schlegel (1968), 176–180.

Partial effect → total cause → past (records)
Partial cause ↛ total effect ↛ future
Total cause = common cause of multiple effects. (Reichenbach 1956, 180)

Key to symbols: → (inference permitted); ↛ (inference not permitted)

Fig. 1. Partial effects and total causes.

more information we cannot predict that it will hit the middle of the pond, that it will cause divergent waves, that it will startle a duck that flies off into the sight of a hunter's rifle. Using these notions, Lockwood attempts to explain the past-future asymmetry. He holds that "events frequently have highly localized nasic conditions in the past" but they do not tend to have "highly localized nasic conditions in the future" (2006, 253). Lockwood's view on the difference between past and future can be summarized as follows:

a. Past outcomes are genuinely overdetermined by their currently prevailing partial effects. The totality of various partial effects overdetermines the total cause. Hence, in line with Reichenbach's conclusion, we can infer the total cause from the various partial effects. Note that we have to learn a lot about the partial effects to be able to exclude alternative causes and ensure that the total cause identified has a much higher probability of explaining the effect than alternatives causes. (Such painstaking detective work went into the identification of the cause of the dinosaurs' extinction sixty-five million years ago. According to the Alvarez hypothesis, it was due to an impact from outer space rather than volcanic activity; but it has not yet been established whether the impact was due to an asteroid or a comet (1997).
b. By contrast, events do not have highly localized nasic conditions in their future (Lockwood 2006, 253). While effects radiate out from some highly localized causal event, it is not the case that some future events are so highly localized that they could serve as nasic conditions for present events. This fact has been labelled the "law of conditional independence": incoming influences emanating from different directions in space are uncorrelated (Penrose and Percival 1962; cf. Price 1996, 118). Consider again Popper's pond analogy to illustrate this difference. When a stone is thrown into the middle of the pond, its waves will diverge towards the shore. From the divergent waves we can infer at least that some local disturbance caused the ripples on the lake's surface, although further investigation may

be required to determine the nature of the disturbance. By contrast, there are no highly localized nasic conditions on the shore of the lake that would cause convergent waves towards the centre. That is, we do not expect a disturbance in the middle of the lake to be caused by correlated conditions on the shore of the lake. It is highly unlikely that conditions on the lakeshore will conspire to produce converging waves towards the centre sufficient to lift a stone from the bottom of the lake. Thus Lockwood's explanation of the past-future asymmetry boils down to the statement that present events are conditionally dependent on past events, while present events are conditionally independent of future events.

But seen in this light, Lockwood's analysis is more like a redescription than an explanation of the asymmetry between past and future events. We can always ask in the case of a physical mechanism why "nasic conditions" are typically highly concentrated in the past and why future actions cannot be regarded as nasic conditions for current events.

2. David Albert introduces a distinction between the prediction of future states and the retrodiction of past states and distinguishes retrodictions from records of past events (2000, Ch. 6). Both predictions and retrodictions are inferences from the present state of affairs to either a future state of affairs (say, the occurrence of a solar eclipse) or to a past state of affairs (say, the identification of an unidentified celestial object in Galileo's notebook as the planet Neptune). The ability to make predictions and retrodictions depends on the use of appropriate equations of motion and the availability of boundary conditions. But according to Albert most of our knowledge about the past cannot be derived from retrodictions; we know it by means of indelible records. Such past records are irreversible data of past events. Albert argues that we have epistemological access to the past other than by means of retrodiction because records are related to a certain feature of the Second Law of Thermodynamics. This feature involves the so-called Past Hypothesis, which states that "the world first came into being in whatever particular low-entropy highly condensed big-bang sort of macro-condition" (Albert 2000, 96; see also Novikov 1998, 204ff). Albert's thesis is that we have records of the past because our experience is "confirmatory of the past hypothesis but not of any future one"[6] (2000, 118).

[6] It should be mentioned in this connection that both Lockwood's assumption that future events are uncorrelated with respect to the present and Albert's assumption that the

It seems implausible to relate the existence of past records to the beginning of the universe and its particular energy state. Rather, as Murray Gell-Mann and James Hartle observe, "a record is a present alternative, that is, with high probability, correlated with an alternative in the past" (1990, §10). The current state of the world is branch-dependent on the past state, where the past state may have happened quite recently: it is "contingent on events that have happened" (Gell-Mann and Hartle 1993, 3346). Lockwood seems to be closer to the mark with his suggestion that nasic conditions are highly concentrated in the past but not in the future. But Lockwood redescribes the asymmetry rather than explaining it. By contrast Albert hints at a connection between the past and the Second Law of Thermodynamics but implausibly explains our experience of the past as confirmatory of a low-entropy Big Bang.[7]

3. Finally, Storrs McCall defends the view that the difference between past and future is that the future comprises many possibilities, while the past is actual and unique (1994, Ch. 1, 2). His explanation is not based on whether we can travel to the past, or whether we can change the past, or whether entropy increases, though it does imply that we cannot change the past. His model is incompatible with the static block universe since if the world is indeterministic (or probabilistic) this view requires that every physically possible future be situated on a different 4-dimensional branch, and that the probability of any future event is specified by the proportion of future branches on which that event occurs. The model is also incompatible with the view that time flow is a subjective phenomenon. The model is in the shape of a tree, the past being a single trunk, the future a multiplicity of branches, and the present the first branch point (Figure 2). Time flow is represented by the vanishing or "falling off" of branches, the one branch remaining being the "actual" one that becomes part of the trunk. The present moves stochastically up the tree. It would continue to do so even if all conscious beings in the universe died out. This view attributes the evo-

entropy of the Big Bang is lower than the entropy of the Big Crunch have been questioned in the literature (see Price 1996 and 2002; Schulman 1997).

[7] It is now commonly assumed that our universe started in a low-entropy Big Bang and has expanded ever since. This expansion, which actually accelerates, as well as the effects of gravity, establishes a connection between the thermodynamic and cosmological arrows of time. Many cosmologists now believe that the universe will end in what the nineteenth century termed a "heat death," that is, a total dissipation of energy that will make life impossible (see Carroll 2009; Penrose 2010). In this sense, there is a connection between the expansion of space and the spatial spreading of energy states.

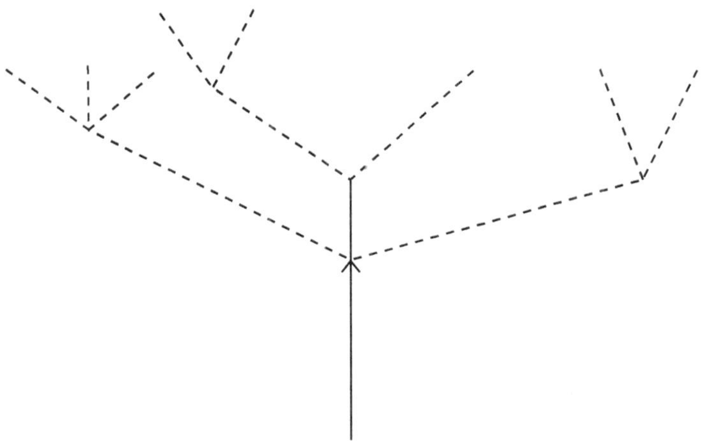

Fig. 2. Branch attrition in the branching tree model, indicating the probabilities of the occurrence of future branches.

lution of the universe, from past to future, from possibility to actuality, to the process of "branch attrition."

This tree model must be compatible with the principle of the Special Theory of Relativity:

> Since in the tree model the future is branched while the past is single, the way in which the universe tree branches will have to be relative not only to a time but also to a frame of reference. (McCall 1994, 10)

As McCall continues, the branched model makes each instantaneous state of the universe

> relative to a frame-time, i.e. relative both to a time and to a coordinate frame.... Becoming, then, is represented in the model by branch attrition within every frame-dependent or hyperplane-dependent branched structure. This does not make becoming observer-dependent, but it does make it frame-dependent. Time flow is not absolute, but is relative to a reference frame. (1994, 32–33)

In the tree model, branch attrition seems to describe the past-future asymmetry in an objective, observer-independent way. But branch attrition is not a known physical mechanism. The model only serves a descriptive purpose. It describes metaphorically why the frozen past is different from the potentialities of the future, but it does not *explain* the past-future asymmetry. Why, for instance, can we not descend down the tree trunk and return to the past? What is needed to make McCall's tree model plausible is a known physical mechanism that can account for "branch attrition." Such a mechanism was also the missing element in Lockwood's redescription. In the recent physical literature the notion of decoherence has been proposed as a mechanism to explain the emergence of quasiclassical sets of histories, i.e. the individual histories of our classical world, which obey with high probability, "effective classical equations of motion interrupted continually by small fluctuations and occasionally by large ones" (Gell-Mann and Hartle 1993, 3345, 3376). Decoherence signifies the emergence of classical macro-states from their underlying quantum states as a result of measurements by their environments. Decoherence leads to different alternative histories for the universe, to branch-dependence of histories and the permanence of the past (Gell-Mann and Hartle 1993, §10). Branch-dependence means that individual histories are "contingent on which of many possible histories have happened" (Gell-Mann and Hartle 1993, 2246–2247). The permanence of the past expresses the feature

> of a quasi-classical domain that what has happened in the past is independent of any information expressed by a future projection. Neither the decoherence of past alternatives nor the selection of a particular past alternative is threatened by new information. (Gell-Mann and Hartle 1993, 3354)

Since decoherence can be understood as the carrying away of phase information into the environment, leading to noise (Gell-Mann and Hartle 1993, 3364, 3376), and as a form of continuous measurement of quantum systems by the environment, leading to entropy (Schlosshauer 2008, 41), decoherence leads to irreversible past records. But irreversibility is a feature of the Second Law of Thermodynamics; hence it is appropriate to associate the past-future asymmetry with the increase of entropy.[8]

[8] Although the Second Law of Thermodynamics was discovered in the middle of the nineteenth century, it is a minor scandal in physics that no unanimous agreement exists on its precise meaning. For the purpose of the following discussion, which uses the notion of entropy to discuss a mechanism for the past-future asymmetry, it will be sufficient to use the established sense of the notion of entropy as it is discussed in statistical mechanics. As human awareness of the past-future asymmetry concerns macroscopic systems, a quantum-mechanical version of entropy that refers to quantum states (see note 10 below) appears to be less useful than the statistical version of entropy. Furthermore, the notion of entropy can be used either to describe the asymmetries in the familiar environment—in which case we should speak of the *passage* of time—or in a cosmological context, in which case we should speak of the *arrow* of time. In the cosmological context, entropy is used to consider both the origin of time's arrow and the ultimate fate of the universe. The universe itself can be understood either as 'our' universe or as the 'multiverse,' of which our universe would be only one universe in many parallel universes (cf. Carroll 2008; 2010; Krauss and Scherrer 2008). Although the arrow of time and its origin are chal-

V. *Entropy*

Note that all of the approaches considered so far are engaged in obtaining a clearer conceptual understanding of the past-future asymmetry, but they still fail to provide a dynamic explanation. It is characteristic of philosophical discussions to consider the conceptual possibility of, say, visiting the past without due attention to physical possibilities. The latter are regarded as merely pragmatic limitations. But as we have seen in the case of Lewis's time traveller, physical considerations can have serious conceptual consequences for the argument at hand. Albert alluded to a dynamic reason for the past-future asymmetry in terms of the Second Law of Thermodynamics, while decoherence encodes a physical mechanism to obtain the classical domain from the quantum realm.

This section of the paper will in conclusion consider a dynamic reason for the past-future asymmetry, one that expands on the earlier criticism of Lewis's time travelling scenario and makes full use of the Second Law of Thermodynamics and the notion of entropy increase. It leads to a general thesis: the entropy of past states is fixed and we cannot change them, but we can influence the entropy of future states.[9] As mentioned above, the

lenging tasks for cosmologists, the present discussion will be limited to the passage of time in our familiar environment because this familiar environment challenges us with the past-future asymmetry. It should be noted that our awareness of asymmetric phenomena is compatible with different topologies of time. For instance, the current understanding in cosmology is that the universe will expand forever, while in fact it is accelerating towards its ultimate fate that the nineteenth century termed "heat death"—the total dissipation of all energy gradients. This scenario suggests a linear topology, that our universe had its beginning in a big bang event but will expand forever. For the inhabitants of such a universe the asymmetry of events, the passage of time, would be the familiar experience. But if the universe, our universe, expanded from a big bang event to a maximum point of expansion and then began to re-contract to end in a big crunch (which may or may not be similar to the big bang event in terms of entropy), the inhabitants of such a universe would still experience the passage of time as asymmetric. The topology of such a universe would be a closed circle—a closed time-like curve—but every section of it would display the familiar asymmetry of our experience. For this reason the entropy notion of the discussion in section V is restricted to the passage of time. The difference drawn here between the local passage of time and the global arrow of time can be described as that between an asymmetry in local regions of space-time and the whole of space-time (Davies 1974, §2.1).

[9] In the present article, entropy is taken to be an important feature of physical systems, in line with classic statements in Davies (1974) and Penrose (2005; 2010). We also assume the validity of the notion of entropy as defined in statistical mechanics and as an accepted part of standard textbooks. For instance, Carroll (2010, 32) calls it Nature's most reliable law. It should be noted that entropy is difficult to measure experimentally and that it is actually inferred from macroscopic parameters, like temperature, pressure, and work (Čápek/Sheehan 2005, 26). However, one cannot pretend that 'entropy' is an uncontested notion. There exists, for instance, a variety of different definitions of entropy in the literature;

discussion will be confined here to classical notions of entropy because humans experience the past-future asymmetry in the macro-world.

It is customary in standard textbooks to describe the increase in entropy in terms of an increase in disorder or loss of information. For instance, the physicist Kip Thorne (1994, 424) illustrates the notion by the case of a father who arranges the toys of his child's playroom in such a way that all the toys are stacked in one corner of the room. It is clear where the toys are. Then the child comes to play and scatters the toys all over the room. There is loss of information as to the whereabouts of the toys. When the child has grown tired and is put to bed, the father returns and restores the toys to their former ordered state. The information is restored. According to the Second Law the overall entropy of the playroom situation, regarded here as a closed system, has increased because the father has to spend energy to restore order in the playroom and the child has spent energy playing (distributing the toys). However, this widespread illustration can also be unhelpful because an increase in entropy can be accompanied by an increase of order. For instance, the overall entropy of the universe since the Big Bang is commonly understood to have increased, but the universe today exhibits a great deal of order as seen in the formation of solar systems, galaxies and clusters of galaxies (Carroll 2008; Davies 1974, §4.6).

A better way to understand entropy is in terms of a *spreading* descriptor: the microstates of a system spread into the available phase space thereby increasing the entropy of the system. To illustrate, consider a sealed container, in which a perfume bottle is placed. There are about 6×10^{20} molecules in this bottle (Figure 3). When the lid is removed from the bottle, the molecules will spread and mix with the air molecules. The amount of spreading is a function of time, since the perfume molecules begin to occupy the space in the container. If we take a snapshot of the spreading at time t_1 ($t_1>t_0$), we observe a certain amount of spreading; at a later time, t_2 ($t_2>t_1$), the spreading will have increased. This spreading can

and the Second Law can be stated without reference to entropy, since the original Clausius statement explains that all work can be transformed into heat but not all heat into work (cf. Čápek/Sheehan 2005, Ch. 1; Uffink 2001; Leff 2007). Furthermore, while the Second Law lacks a thorough theoretical proof, "its empirical support is vast and presently uncracked" (Čápek/Sheehan 2005, 42; Sklar 1993, Ch. 7). Although its theoretical foundation is uncertain, it rests on firm empirical evidence, which justifies its use in the present argument. Despite the disagreement about the theoretical grounding of entropy, it will be assumed that a spreading descriptor (Leff 2007, 1744) is an appropriate tool to support the argument in favour of the past-future asymmetry.

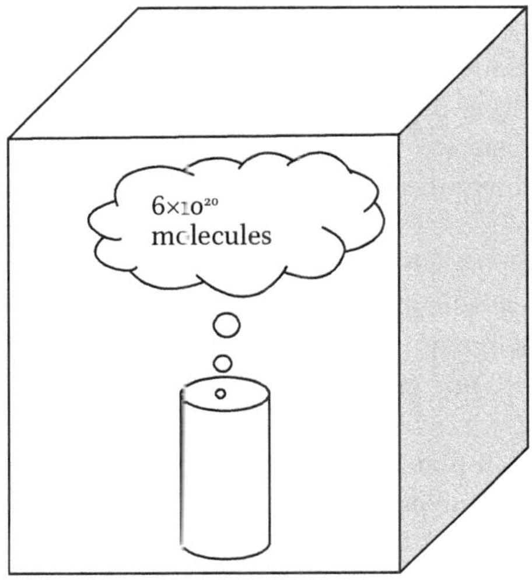

Fig. 3. Expanding molecule cloud in phase space; spreading as a function of time.

serve as a primitive clock. Statistically speaking the amount of spreading will increase with high probability until the system reaches equilibrium. We can only stop the spreading if we decide to interfere with the system from outside, for instance by partitioning the box before the perfume molecules have filled all the available space. Then the spreading and increase in entropy are stopped by the injection of external energy.

This spreading metaphor may be made more precise in the following manner:

> An increase of entropy may be said to correspond to a "spreading" of the system over a large number, *W*, of occupied quantum states. Alternatively one might say that entropy is a measure of the extent to which the system in question is unrestrained; the less constrained it is, the greater the number of its accessible quantum states for given values of those constraints which exist. (Denbigh/Denbigh 1985, 44; cf. Čápek/Sheehan 2005, 17)[10]

---

10 The quote explains spreading in terms of accessible quantum states. In quantum mechanics, entropy is a state defined via the density matrix, ρ: $S_{\vartheta N}(\rho) = -kTr[\rho\, ln(\rho)]$, where the left-hand side of the equation is a measure of "the quantity of chaos in a quantum mechanical mixed state" (Čápek/Sheehan 2005, 17). Although the quantum mechanical definition is more basic, the past-future asymmetry is concerned with the macroscopic experience of human agents, which justifies the restriction of the discussion to a classical understanding of entropy.

Here *W* is the thermodynamic probability, which is related to entropy by the Boltzmann relation: $S=k\ln W$. In order to appreciate this relation—actually the definition of entropy in statistical mechanics—we must distinguish the macrostates of a system, like pressure and temperature, from its microstates, which make up the constellation of the microscopic constituents of the system. The molecules from the perfume bottle are the microscopic elements, while the pressure of the molecules on the container wall is a macroscopic state. It generally holds that any macrostate corresponds to a large number of microstates, but that any given microstate corresponds to one given macrostate. A useful analogy is to think of the number of ways in which you can pay for an item that costs say, £10. You may tender a ten-pound note, ten one pound coins, twenty-five pence coins, one thousand one pence coins, etc. The coins and notes correspond to the microstates, whilst the value of £10 corresponds to the macrostate. Thus any given macrostate in a physical system, like pressure or temperature, can be realized by a certain number of microstates. *W* is then the number of ways in which a macrostate can be realized. If $W = 1$ the macrostate is in a state of least spreading, but as *W* increases the spreading of the microstates increases, and so does entropy. The spreading of the energy states of the system is at all times accompanied by a distribution of the energy states in a phase space. More precisely, *W* is related to the number of arrangements in Γ-space, a six-dimensional space, which specifies the microscopic configuration *(qi, pi)* of an N-particle system. This description is needed to account for the motions of the individual molecules, in terms of momentum. Although it is customary to describe the increase in entropy as an increase in disorder, it is more appropriate to think of the increase in entropy in terms of the increase of *W*:

> A single system's phase point traverses the phase space in time as particles move and exchange energy. This traversal provides a graphic image of a system's particles spreading and exchanging energy through space, as time progresses. Energy, space, and time are all explicitly involved. In this sense, entropy can be thought of as a *spreading* function. (Leff 2007, 1750; italics in original)

To explain this spreading function further, both the spatial spreading of processes and the temporal spreading of system states over accessible states must be considered (Leff 2007). For instance, in the case of the child's playroom the spreading of the toys resembles a spatial disintegration of a neat pile, but spatial spreading makes no explicit reference to energy balances. By contrast temporal spreading refers to the expansion of system states into the available phase space, as expressed in Boltzmann's statistical definition of entropy.[11] Boltzmann's statistical definition refers directly to the "colonization" of the available phase space by the microstates, since the Boltzmann entropy is a measure of the number of microstates that are compatible with a given macrostate. Even though the equations of motion are time-reversible invariant, the spreading, say, of gas molecules into the available phase space occurs because there are many more ways of populating a larger than a smaller space. Another way of saying this is that, given the spread of the later state, the probability of the coherence of final conditions that would be required for the temporal inverse of typical trajectories is so vanishingly small that their physical realization is negligible. This argument appeals to the differences in the topology of the phase space of a system under consideration (Hill/Grünbaum 1957).

The idea of the spreading of energy states can be applied to the past-future asymmetry. Any past moment corresponds to an amount of spreading over the available phase space, but the present moment from which the past is observed corresponds to a greater amount of spreading. As the present is an open system, we cannot manipulate the ensemble of microstates to return them to their past constellation.

As we have seen, records are evidence of the branch-dependence of current states on past states. If past records are considered as decohered states—i.e. the result of an interaction with their environment, technically described as a loss of interference terms—then decoherence leads to an

---

11 In order to emphasize the importance of the spreading function, reference should be made to Liouville's theorem, which states that the temporal evolution of classical dynamic systems preserves volumes of phase space regions but not the shapes of these regions. A rather uniform region in the initial stage can become very fibrillated after the spreading of the system. Liouville's theorem in classical mechanics states that a volume element along a flowline conserves the classical distribution function, $f(r,v)\ drdv$: $f(t + dt, r + dr, v = dv) = f(t, r, v)$ (Kittel/Kroemer 1980, 408; Albert 2000, 73f). In other words, if we consider trajectories in phase space, which include both position and momentum of particles, then the equation of motion of such systems can be expressed in terms of its Hamiltonian, *H*. *H* expresses the conservation of total energy of the system. Liouville's theorem then states that the volume of the phase space, which an ensemble of trajectories occupies, remains constant over time. Liouville's theorem shows that the volume of the phase space regions is invariant over time even though the expansion of the trajectories within this volume can start from different initial states or end in different final states. But an immediate consequence of this theorem is that even though the *volume* is preserved the *shape* of this phase space region is not preserved and this implies a dynamic evolution of the trajectories within this region. For two shapes cannot differ from each other without an evolution of the trajectories. It also implies that a reversed evolution of the trajectories will preserve the volume but not the shape and hence that reversed trajectories need not be invariant with respect to the shape of the phase space region.

increase in entropy. Hence the time traveller, considered above, cannot decohere past records. Records reflect a differential in energy states between the present and the past. Records are evidence of past entropy states, but these are not necessarily ordered states—like a well-preserved fossil—but can also be disordered states—like a decomposed organism. One immediate problem with this suggestion is that, on purely statistical grounds, records of the past are as likely to have emerged from a higher entropy state by a fluke fluctuation as from a lower energy state.[12] (cf. Weizsäcker 1937; Earman 1974, §7; Horwich 1987, Ch. 4; Carroll 2010, Ch. 9.) Earlier, we referred to the notion of decoherence, which is an umbrella term describing various physical mechanisms that lead to different alternative histories and branch-dependence. In order to deal with the objection just mentioned, it is important to recall the fundamental fact that in the theory of decoherence it is normally not the case that every history of a system can be assigned a probability (cf. Craig/Singh 2011). In other words, some possible histories have a negligibly small probability of occurring, while others, like the familiar histories of classical systems, have a high probability. Thus, while all histories are equally possible, they are not all equally probable. As shown below, most systems have a negligibly small probability of returning to their initial states. The probability of a history depends on the level of its entanglement with other systems that act as a measuring environment. And decoherence needs to be paid for by an increase in entropy (Schlosshauer 2008, §2.8).

Why concentrate on entropy? First, it is a universal relation in the sense that no empirical violation of the Second Law is known (Čápek/Sheehan 2005, 13, 42; Carroll 2010, 284). Even though it is only statistical in nature, according to statistical mechanics, it affects many other arrows of time. Second, it provides a dynamic explanation of the past-future asymmetry that was missing from other accounts. The fact that the Second Law is statistical in nature may invite certain reservations. It means that, in principle, a state in higher entropy may spontaneously return to a state

[12] The temporal neutrality of statistical relations also throws up the notorious problem of Boltzmann brains: thermal fluctuations may give rise to human brains and their complexity in an otherwise high-entropy universe. The universe surrounding human observers is manifestly not in a state of high-entropy, and if it is the result of a fluctuation it was not a random fluctuation, if Carroll's scenario of baby universes is accepted (see Carroll 2008; 2010, 221–224, 355–359). In a different context, Horwich makes the relevant suggestion that the "time reverse of every possible process is equally possible, though not equally probable" (1987, 54).

of lower entropy, and such an unobserved reversed process is compatible with the Second Law. For instance, a broken cup may spontaneously reassemble itself. In fact, according to Poincaré's theorem a finite mechanical system, whose state S_0 is characterized by the position (q_k) and momentum (p_k) variables of its micro-constituents, will return as closely as possible to the initial set of variables in Poincaré recurrence time (1890). Thus the Second Law, in its statistical version, gives rise to *de facto* (not *de jure*) irreversibility.

The importance of *de facto* irreversible processes in the past-future asymmetry has been emphasized by H. Reichenbach and A. Grünbaum. *De facto* irreversible processes are to be understood as complex processes whose time-reverse is highly improbable in the history of the entire universe, although they remain theoretically possible. If they occurred, they would not violate the laws of the micro-processes in terms in which the macro-processes are understood. The fact that the reversibility of physical processes, if it occurs, is highly unlikely and does not violate the Second Law, has been called weak T-invariance (Grünbaum's notion of *de facto* irreversibility).

This weak T-invariance must satisfy the "requirement that its time inverse (although perhaps improbable) does not violate the laws of the most elementary processes in terms of which it is understood" (Landsberg 1982, 8). This take on things implies that the T-invariance of physical laws is compatible with asymmetric solutions, if appropriate boundary conditions are taken into consideration (Price 1996, 88–89, 96; Denbigh 1981, Ch. 6.2). Thus, it is not a violation of the Second Law that a cold cup of coffee spontaneously reheats itself at some stage in the future history of the universe—through a fortunate self-rearrangement of the molecules—but such behaviour has never been observed. Nor is it expected to be observed in the remaining course of the history of the universe. But how improbable is the reversal of such processes? One aspect of an answer to this question is that the Poincaré recurrence time only exists for isolated systems in classical physics, but modern quantum physics emphasizes the importance of open systems:

> Another version of the same argument is well known: why not reverse all the molecules' velocities so the system will go back to its initial state? The answer is again probabilistic among all possible velocity distributions, the ones returning to the initial state have a negligible weight. (Omnès 1999, 239)

This argument relates the probability of occurrences to the available number of realizable states, i.e. the inequality in the topology of the initial and final states. If W is much greater for the later state, it reduces the probability of the coherence of final conditions to return the system to its initial lower-W state.

To illustrate the negligibly small probability of the recurrence of a system to its initial state, note that the time scale for a Poincaré recurrence is of the order of $10^{10^{25}}$ years for a gram mole (Denbigh/Denbigh 1985, 140; Ambegaokar/Clerk 1999). Or to give an example on the cosmic scale, cosmologists estimate the amount of time it would take for a volume of gas containing 10^{18} molecules to return to its initial state (position and momentum variables). It is assumed that each molecule (with an average molecular velocity of 5×10^{14} cm/s in both directions) would return to within 10^{-7} cm of each initial position variable and within 10^2 cm/s of each velocity variable. The estimated time for a return to such a configuration would require $10^{10^{19}}$ years, which is well beyond the estimated age of the universe (~10^9 years) (see Schlegel 1968, 52–53). It has long been recognized that an increase in entropy should not be used to identify the arrow of time, precisely because of the theoretical possibility of recurrence (Schlegel 1966). Although we have *de facto* irreversibility, we have *de jure* reversibility of physical systems that obey the fundamental equations of motion both in classical and quantum physics. It remains nevertheless true that *de facto* irreversible processes are so overwhelmingly probable that they can serve as a dynamical explanation of the past-future asymmetry in our familiar world.

If the increase in entropy is a *de facto* irreversible process, should arrows of time be linked to entropy increase? Would the arrow of time not reverse if a given system implausibly returned to its initial state? The answer adopted here is that it would be a mistake to identify arrows of time with entropy increases because of the T-invariance (reversibility) of fundamental equations and the temporal neutrality of statistical relations. Several writers have warned against "taking thermodynamics too seriously" (Callender 2001; Uffink 2001; Horwich 1987; Earman 1974). But this recognition does not prevent us from regarding entropy—or more precisely, the thermodynamic probability W—as one of the *criteria* from which the anisotropy of physical time can be inferred. There are many parallel processes in the universe—a rise in the entropy gradient of many systems, the cosmological expansion of the universe, the emergence of classical systems through decoherence mechanisms—all indicating a past-future distinction. In addition, it must be recognized, as Popper's lake analogy illustrates, that the boundary conditions of the universe—in the present case the low-entropy initial conditions of our universe—must be taken into account when we seek to account for the arrows of time. A complaint of proponents of temporal symmetry against this approach has been that the boundary conditions are put in 'by hand.' But this objection has lost a lot of its force since present-day cosmology is precisely concerned with explaining events like the Big Bang and its low entropy initial state (cf. Gott 2001; Carroll 2010; Penrose 2010). This research also strongly suggests that there is no evidence of a future low-entropy state of the universe. Even proponents of temporal symmetry concede that t-symmetric laws may have t-asymmetric solutions (Earman 1967, 548; Price 1996, 88). When all these processes are taken into account, we possess reliable criteria from which to infer the anisotropy of time.

VI. *Conclusion*

On this entropy account, then, we can explain several aspects of the past-future asymmetry:

1. We cannot travel into the past in ordinary time machines because even as non-participant observers we would interfere with the entropy balance of the past, which is fixed. In other words, time travellers cannot manipulate the conditions that would allow them to roll back the amount of spreading that has taken place between, say, 1921 and 1957. At the time of grandfather's youth this distribution of the phase space had reached a certain configuration: going back to the past, even as an innocent bystander, would necessarily change this configuration. This cannot be done because what separates grandfather's configuration of the phase space from that of the time traveller's—his grandson—is a change in the distribution of states. One could say that the change in the distribution of states—for which grandfather was causally responsible—made grandson possible. The trajectory leading from grandfather to grandson is not a spatial trajectory, on which we could travel up and down a certain number of times. This is Wells's mistake. It is a spreading of energy states, which cannot be undone, because the energy spent on the trajectory has changed the entropy balance of the states and this state cannot be regained by the input of new energy. The spreading is also sequential. The time traveller cannot weave his way back through the multiple sequences of energy states because none of the energy spent between grandfather's time and grandson's time is recoverable, since only part of this energy will have been available to do useful work

(as is clear from the First Law of Thermodynamics). But even if it were possible for a Maxwellian demon to reverse the spreading and restore a former energy state, the demon's actions would not constitute a return to the past but a return to a copy of the past. It would reemphasize the essentially linear nature of temporal processes in the natural world.

2. While the state of spreading of the past cannot be changed, we can influence the future, as illustrated in the branching tree model. From the point of view of the present state of the universe, we can change the spreading of states into the future because we are free to use the available energy today to channel it in particular directions. If we decide to make a cup of coffee we use the available energy; if we do not, the energy remains available to do other useful work. We can choose not to open the perfume bottle or to stop its spreading by inserting a partition in the container.
3. We have records of the past but not of the future because records are manifestations of the entropic state of the world of the past. Records may be either of ordered systems—like footprints on a beach—or of disordered systems—like a dilapidated building. When Lockwood stresses the concentration of nasic conditions in the past, for which we have no equivalent in the future, he presumably has such records in mind. We do not have such records of future states because a) present spreading is contingent on past but not on future records; and b) spreading is a progressive process contingent on the past history of the system.

The past-future asymmetry has thus been grounded in a dynamic explanation that refers to the Second Law of Thermodynamics. According to the entropy view, the past-future asymmetry exists objectively in the physical world because the energy balance of past stages of the world exists at a different level of 'spreading' than present and future energy balances.

References Cited

Albert, David. 2000. *Time and Chance*. Cambridge (Mass.)/London: Harvard UP.

Alvarez, Walter. 1997. *T.rex and the Crater of Doom*. Princeton (New Jersey): Princeton University Press.

Ambegaokar, Viney and Aashish A. Clerk. 1999. "Entropy and Time," *American Journal of Physics* 67/12: 1068–1073.

Atkins, Peter. 2003. *Galileo's Finger*. Oxford: OUP.

Boltzmann, Ludwig. 1878. "Weitere Bemerkungen über einige Probleme der mechanischen Wärmelehre," *Wiener Berichte* II: 7–46.

Born, Max. 1962. *Einstein's Theory of Relativity*. New York: Dover.
Brown, Harvey. 2005. *Physical Relativity*. Oxford: Oxford University Press.
Čápek, Vladislav and Daniel P. Sheehan. 2005. *Challenges to the Second Law of Thermodynamics*. Dordrecht: Springer.
Callender, Craig. 2001. "Taking Thermodynamics too Seriously." *Studies in History and Philosophy of Modern Physics* 32: 539–553.
Carroll, Sean. 2008. "The Cosmic Origins of Time's Arrow." *Scientific American* (June): 48–57.
——. 2010. *From Eternity to Here*. Oxford: One World.
Craig, David and Parampreet Singh. 2011. "Consistent Histories in Quantum Cosmology." *Foundations of Physics* 41: 371–379
Davies, Paul. 1974. *The Physics of Time Asymmetry*. Leighton Buzzard/London: Surrey University Press.
Davies, Paul. 2001. *How to Build a Time Machine*. London: Allen Lane, The Penguin Press.
Denbigh, Kenneth George. 1981. *Three Concepts of Time*. Heidelberg/London/New York: Springer.
Denbigh, Kenneth George and Jonathan Stafard Denbigh. 1985. *Entropy in Relation to Incomplete Knowledge*. Cambridge: Cambridge University Press.
Dummett, Michael. 1986. "Causal Loops." In *The Nature of Time*, edited by Raymond Flood and Michael Lockwood, 135–169. Oxford: Blackwell 1986.
Earman, John. 1967. "Irreversibility and Temporal Asymmetry." *The Journal of Philosophy* 64: 543–549.
——. 1974. "An Attempt to Add a Little Direction to 'the Problem of the Direction of Time.'" *Philosophy of Science* 41 (1): 15–47.
——. 2006. "The 'Past Hypothesis': Not even false." *Studies in History and Philosophy of Modern Physics* 37: 399–430.
Ehlers, Jürgen. 1997. "Concepts of Time in Classical Physics." In *Time, Now, Temporality*, edited by Harald Atmanspacher and Eva Ruhnau, 191–200. Heidelberg/London/New York: Springer.
Gell-Mann, Murray and James B. Hartle. 1990. 'Quantum Mechanics in the Light of Quantum Cosmology." In *Complexity, Entropy and the Physics of Information*, edited by Wojciech Zurek, 425–465. Boston (Mass.): Addison-Wesley.
——. 1993. "Classical equations for quantum systems." *Physical Review* D 47/8: 3345–3382.
Geroch, Robert. 1978. *General Relativity from A to Z*. Chicago/London: The University of Chicago Press.
Gott, Richard. 2001. *Time Travel in Einstein's Universe*. London: Phoenix.
Grünbaum, Adolf. 1955. "Time and Entropy." *American Scientist* 43: 550–572.
——. 1967. "The Anisotropy of Time." In *The Nature of Time*, edited by Thomas Gold, 149–177. Ithaca: Cornell University Press.
Hill, Edward L. and Adolf Grünbaum. 1957. "Irreversible Processes in Physical Theory." *Nature* 179, 1296–1297.
Horwich, Paul. 1987. *Asymmetries in Time*. Boston: MIT.
Janssen, Michael. 2009. "Drawing the line between kinematics and dynamics in special relativity." *Studies in History and Philosophy of Modern Physics* 40: 26–52.
Kant, Immanuel. 1787/1976. *Critique of Pure Reason*, Trans. N. Kemp Smith. London: The Macmillan Press.
Kittel, Charles and Herbert Kroemer. 1980. Second Edition. *Thermal Physics*. New York: W.H. Freeman and Company.
Krauss, Laurence M./Robert J. Scherrer. 2008. "The end of cosmology." *Scientific American* (March): 47–53.
Landsberg, Peter T. 1982. "Introduction." In *The Enigma of Time*, edited by Peter T. Landsberg, 1–30. Bristol: Adam Hilger.
Leff, Harvey S. 2007. "Entropy, Its Language and Interpretation." *Foundations of Physics* 37: 1744–1766.

Lewis, David. 1976. "The Paradoxes of Time Travel." *American Philosophical Quarterly* 13: 145–152.
Lockwood, Michael. 2005. *The Labyrinth of Time*. Oxford: Oxford University Press.
Lucas, John R. 1973. *A Treatise on Time and Space*. London: Methuen.
McCall, Storrs. 1994. *A Model of the Universe*. Oxford: Clarendon Press.
Nahin, Paul. 1993. *Time Machines*. New York: Springer.
Norton, John. 2010. "Time Really Passes." *Humana.Mente* 13: 23–34.
Novikov, Igor D. 1998. *The River of Time*. Cambridge: Cambridge University Press.
Omnès, Roland. 1999. *Understanding Quantum Mechanics*. Princeton: Princeton University Press.
Penrose, Oliver and Ian C. Percival. 1962. "The Direction of Time." *Proceedings of the Physical Society*. 79: 605–616.
Penrose, Roger. 2005. *The Road to Reality*. London: Vintage Books.
——. 2010. *Cycles of Time*. London: The Bodley Head.
Poincaré, Henri, J. 1890. "Sur le problème des trois corps et les équations de la dynamique." *Acta Mathematica* 13: 1–270.
Popper, Karl. 1956. "The Arrow of Time." *Nature* 177: 538.
Price, Huw. 1996. *Time's Arrow and Archimedes' Point*. Oxford: Oxford University Press.
——. 2002. "Burbury's Last Case: The Mystery of the Entropic Arrow." In *Time, Reality and Experience*, edited by Craig Callender, 19–56. Cambridge: Cambridge University Press.
Reichenbach, Hans. 1956. *The Direction of Time*. Berkeley: University of California Press.
Schlegel, Richard. 1966. "Time and Thermodynamics." In *Voices of Time*, edited by Julius T. Fraser, 500–526. New York: Braziller.
——. 1968. *Time and the Physical World*. Mineola (NY): Dover Publications.
Schlosshauer, Maximilian. 2008. *Decoherence and the Quantum-to-Classical Transition*. Heidelberg/London/New York: Springer.
Schmidt, Brain P. 2005. "The Expanding and Accelerating Universe." In *Cosmic Explosions*, 511–523. Springer Proceedings in Physics, 2005, Volume 99, Part VIII.
Schulman, Laurence. 1997. *Time's Arrow and Quantum Measurement*. Cambridge: Cambridge University Press.
Sklar, Laurence. 1993. *Physics and Chance*. Cambridge: Cambridge University Press.
Thorne, Kip S. 1995. *Black Holes and Time Warps*. London: Papermac.
Uffink, Jos. 2001. "Bluff Your Way in the Second Law of Thermodynamics." *Studies in History and Philosophy of Modern Physics* 32: 305–934.
Weinert, Friedel. 2010a. "Relativistic Thermodynamics and the Passage of Time." In *Humana. Mente* 13: 175–191.
——. 2010b. "Minkowski Spacetime and Thermodynamics." In *Space, Time and Spacetime*, edited by Vesselin Petkov, 239–526. Heidelberg, Berlin, New York: Springer 2010 (*Fundamental Theories of Physics* Vol. 167).
Weizsäcker, Christian F. von. 1939. "Der zweite Hauptsatz und der Unterschied von Vergangenheit und Zukunft." In Christian F. von Weizsäcker. 1979. Fifth edition. *Die Einheit der Natur*, II. 2. München: Hanser. Originally published in *Annalen der Physik* 428 (3–4): 275–283.
Wells, Herbert G. 1895/2005. *The Time Machine*. Penguin Classics.
Whitrow, Gerald J. 1980. *The Natural Philosophy of Time*. Oxford: Oxford UP.

SECTION THREE

FUTURES OPENED, FUTURES CLOSED

THE HUMAN TEMPORAL CONDITION BETWEEN MEMORY AND HOPE

Steven Ostovich

Abstract: This essay brings together some works of Hannah Arendt and J.T. Fraser, two thinkers who took seriously our embedded-ness in time, in order to frame a possible understanding of the human temporal condition. Arendt coins the word "natality" to describe freedom as the capacity of human action to give birth to or be the origin of new things in the world. Memory is the basis for the identity and responsibility of the actor. But is memory an adequate response to what Fraser describes as the nootemporal problem of death? Giving direction to action, a nootemporal necessity, is a matter of promising: promises are an indication of an anticipated future and hope. But does promising make sense in the face of the unpredictability of human action? Memory is fragile, and hope seems irrational. In response to this condition this essay will propose an eschatological understanding of time as key to making sense of memory and hope as rational activities.

Keywords: Natality, Hannah Arendt, J.T. Fraser, sociotemporality, evolution of time, *vita activa*, dangerous memories.

What is "the human temporal condition," and what does it mean to describe this condition as situated "between memory and hope"? What follows is an attempt to respond to these two questions. An answer to the first question might seem obvious given the fact that we are the animals who think about time and are aware of the fact that our lives pass in time. And yet it is this awareness of passing that makes it hard for us to take time seriously. We are mortal, and our temporal finiteness tempts us to frame our self-understanding in terms that promise release from our temporal condition and death: we can define our "nature" as an eternal spirit or soul beyond the vicissitudes of time. Another way to state the first question, then, is whether or not we can make sense of our existence in time without this kind of denial of our condition. I hope to show below that this is a practical question of how we should act in light of the awareness of our temporal condition and that this acting should be informed by memory and hope.

In this process of responding to these questions I will be engaging the thoughts of two scholars whose major works reflect in varied ways their concern with topics relevant to answering our questions. The late J.T. Fraser was a time scholar or "time smith" who spent a lifetime developing what he labeled both a hierarchical theory of time and a theory of time's conflicts. In *Time and Time Again: Reports from a Boundary of the Universe,* Fraser gives us a way to think through time without denying time's evolutionary complexity. The other thinker is Hannah Arendt, whose *The Human Condition* is an extended meditation on human existence in view of what she calls "natality" or the human capacity to be the origin of the new. Fraser and Arendt offer us possibilities for understanding our temporal condition as well as guidance for how to live well as beings who exist in time and are aware that they do so; that is, they help us live in memory and hope.

In pairing Fraser and Arendt I am not claiming there is some heretofore unnoticed connection between them. It is rather the case that reading them together might allow us to see new things as they challenge us to work through seemingly disparate concepts like evolution and eschatology when thinking about origins. To be specific, there are two ways in which what follows might be helpful and contribute to the "cash value" of this work. The first has to do with Fraser's hierarchical theory of time/time's conflicts. Fraser identifies five of what he names the "*canonical forms* of time and causation" (91). Listed in order of increasing complexity, these are the "atemporality" of the world of electromagnetic radiation; the "prototemporality" of elementary objects; the "eotemporality" of massive matter; the "biotemporality" of living organisms; and the "nootemporality" of the mature human mind. Each of these forms of time constitutes an "umwelt," and it assumes that "each temporality is complete for the organizational level where it is identified, even if incomplete when compared with the human experience of time" (91). This incompleteness is apparent only from the viewpoint of the "higher" or encompassing level of time in each case, and awareness of this incompleteness arises from the conflicts characteristic of each level.

Fraser also points towards a sixth canonical form of time, "sociotemporality" (47), a temporal umwelt developing in response to the conflicts of nootemporality like the awareness of personal death. Sociotemporality seems to come into being with the evolutionary development of human beings, and we become aware of it in among other places the nootemporal conflict between the freedom of the individual and the freedom of groups to which individuals necessarily belong. Many of us have been speculating about sociotemporality as illustrated by a recent issue of *KronoScope* which includes a report from Albert Mayr on "The Round Table: A Social

Time Sculpture" and a review essay by Mark Aultman on "Nootemporality, Sociotemporality, and the Internet."

What follows may be construed as my attempt to contribute to this discussion of sociotemporality, although sociotemporality will remain in the background. Work in this area so far has been hampered not only by the limits of nootemporality but perhaps by the term itself and the context it implies. *Socio*temporality would naturally be assumed to be the form of temporality characteristic of the umwelt of social groups. Fraser encourages thinking about sociotemporality in this context when he writes, "Societies create those conditions and operational properties of living and thinking humans in collectives that define the social present and allow the emergence of sociotemporality" (47) and "We may talk about *sociotemporal* umwelts for human groups if they are sufficiently coordinated to act in self-interest" (157). Fraser acknowledges, however, that "The epistemic difficulties in identifying the hallmarks of social time are great," and that, "there are reasons to believe that in dealing with societal time we shall need a new language whose vocabulary and syntax derive entirely from social and not from individual experience" (157). Fraser himself only accomplishes a shift from the individual to the group as the subject of nootemporality rather than moving to a new level of temporal complexity.

These "epistemic difficulties" are exacerbated by what appears to be a mistaken contextualization of sociotemporality. If sociotemporality is the time world of groups as opposed to the nootemporal time world of individuals, tensions between individuals and groups become one of the conflicts sociotemporality is supposed to resolve. But Aultman, for example, expresses the concern that if the relationship between nootemporality and sociotemporality is hierarchical, as it must be in Fraser's morphologies, we end up with a problem of "deference to authority" in the conflict between the individual and the group (130–131). Other explorations of sociotemporality start with the classic conviction that to be human is to be a political animal, but this would seem to entail collapsing sociotemporality into nootemporality as forms of human time that evolve together: what, then, is the difference between them? Is it simply a matter of whether one starts with the individual or the group? And wouldn't this confuse the nature of the temporal umwelt with the number of persons involved in it? Also, in Fraser's description of sociotemporality in terms of the "self-interest" of groups there seems to be lurking the kind of teleological goal-orientation that belongs properly to nootemporality in positing what seems to be a group "mind." Identifying sociotemporality as the temporal umwelt of groups is simply too conservative to do justice to Fraser's insight into "the capacity of man to change his social institutions in response to symbolic causes" (158), that is, the human capacity to disrupt time's flow by doing the radically new.

This brings us to the second way in which this paper might be helpful in thinking about the human temporal condition. In the attempt to live up to the radical challenge of sociotemporality in what follows, we will turn to Arendt for orientation in her introducing of "natality" as the human capacity to do the new. But there is also a traditional religious and literary understanding of time that I believe is an undervalued resource for our thinking about time: *eschatology*. In the last section of this essay I will use the "vocabulary and syntax" of eschatology to think some more about human temporal existence. To do so I will have to turn to the biblical apocalyptic literature that is the source for the symbols that inform eschatological thinking. My concern, however, is not spelling out the characteristics of the apocalyptic literature but to have us focus on the nature of eschatological time as *interruption*, and I want to emphasize the difference between eschatological rupture and teleological continuity. This difference is hard to see insofar as eschatology has been "secularized" in the philosophy of history as progressive and oriented towards a promised goal, whether that goal is secured by science or promised by Hegel, Marx, or someone else. As Karl Löwith pointed out, this secularization may be a good thing as it offers the possibility of "taming" the dangerously disruptive element of eschatology, and I must acknowledge our discomfort with eschatology based on our experience with eschatological movements that justify apocalyptic violence through appeals to utopian visions of an ideal future. The history of eschatological violence and oppression cannot be denied. And yet this is not the whole eschatological story. The issue for those who take eschatology seriously is how to be critical as well as eschatological in our thinking. In eschatology we have a way to think through the ruptures that disturb the nootemporal umwelt.

Evolution

It may seem contradictory to turn to evolution immediately after discussing eschatology and temporal rupture given the emphasis of evolutionary biologists on constructing smooth flowing narratives of change. But thinking in evolutionary terms is part of understanding Fraser's hierarchical theory of time's conflicts: for Fraser, the relationship among the canonical forms of time is evolutionary. His claim is that "time had its genesis at the birth of the universe, has been evolving along a scale of qualitative changes appropriate to the complexity of the distinct integrative levels of natural processes and remains evolutionarily open-ended" (48).[1] The direction Fraser sees in temporal evolution also is significant: "the evolutionary journey of man may be described as one from darkness to light but not one from time to the timeless. Rather it is one from the primitive reality of whatever is timeless, to the immense wealth of whatever is temporal" (13). There is no escape from time and the hope that time might empty itself in eternity—and that the sting of death might thereby be removed—is illusory and based on a backwards view of time's arrow. Challenging a fundamental belief of Western thinking, Fraser claims Plato was wrong in his theory of knowledge as movement towards the eternal: "It is the temporal wherein intelligibility and light reside" and "only the temporal is open to creation and freedom. The liberation of the soul is a ceaseless journey from the timeless to the temporal" (17). In the *Republic,* Plato uses the image of a dividing line between this world of mere changing appearance and the eternal, unchanging truth of mathematics, science, and philosophy, and he asks us to cross this line using dialectic to lead us out of time and into truth (Book 6, 508c–511e); similarly, his Allegory of the Cave (Book 7, 514a–519d) describes our epistemological condition in terms of the darkness of a cave and asks us to use reason to throw off the chains that bind us in this world of moving shadows and ascend to the timeless and true light of day (while also imposing the obligation on those who have seen the light to return to darkness to liberate others). What Fraser claims, then, is that Plato's divided line must be inverted and the journey out of the cave and back reversed—truth lies in time.

Time's evolution is towards complexity rather than eternity. And the leading edge of evolution or the boundary of complexity is the human brain. It is "by virtue of possessing human brains" that "we are right at the boundaries of natural complexity" (Fraser 25). Our brains are the source of nootemporality and are what enable us to frame a theory of time's conflicts. And if time is not the "background to reality" but "constitutive of

[1] Fraser, *Time and Time Again*. The quote continues, "Earlier temporalities are not replaced but subsumed by latter ones," a claim reminiscent of Joseph Needham's evolutionary view of biological complexity (see Needham 233–272).

reality" (48)[2] a reality in which we are "active, creative participants" (33), then we are the process of temporal evolution become self-conscious.

Expressed this way, Fraser's insight echoes that of Julian Huxley in *Religion without Revelation* concerning human beings as the pinnacle of creation and the place where evolution becomes aware of itself and therefore free (181–212; ch. 9). But whereas for Huxley this offers the good news that humans might now be able to direct evolution on an orderly course, Fraser remains consistent in believing that evolution leads not to the rest of eternity but to complexity and with complexity, conflict: "A complete view of Creation must insist that what emerged from the primeval, absolute chaos was not order but conflict between orderability and disorderability, and that the history of the universe has been one of the generation of increasingly more sophisticated, new forms of conflict" (95, see also 146). Sounding a little like Trotsky, Fraser describes this condition as permanent revolution and claims, "human freedom is responsible for permanent revolution" (250). Human freedom in complexity and conflict also is the source of human creativity (and vice versa). Fraser describes this creativity in terms of our noetic position in temporal evolution as "descent" and "ascent," that is, the creative act is a descent into earlier forms of temporal consciousness—down into the biotemporal, eotemporal, prototemporal, even atemporal—and a re-ascent through these temporal umwelts to the present. Freeing us from the cares and anxieties of the noetic present, we are liberated into a reality that requires us to employ our freedom in judgment and responsibility. Fraser several times represents this descending/ascending creator motif by pairing the stories of Lot's wife—the descent into the stone-like form of a pillar of salt—and Galatea, the ascent of stone come to life (see 31–33, 175, 204). Physiologically, he identifies this "aesthetic adventure" as the "continuous, self-scanning survey of the brain" that constitutes "consciousness" (175). And he characterizes this experience in Heideggerian philosophical terms as the awakening of a dreamer who is "thrown" into time (189). "Each decision is an act of creation," which explains why our "stunning creativity" is allied with "frightening destructiveness" (169, 267). Fear and hope once again become relevant categories for understanding human existence.

[2] See also: "Received views ... regard time as a background to reality.... At variance with these views, the hierarchical theory of time regards time as constitutive of reality, as a symptom or corollary to the complexity of nature" (Fraser 90).

Origin

"With the creation of man, the principle of beginning came into the world itself, which, of course, is only another way of saying the principle of freedom" (*The Human Condition* 177). With these words Arendt keeps us in the same place as Fraser, a place where freedom is connected to creativity, complexity, and conflict, a place of inventiveness, discovery, and originality. She describes the "character of startling unexpectedness" that is "inherent in all beginnings and all origins" (178), and that makes our actions seem like "miracles." The term Arendt coins for this capacity to give birth to the new is "natality."

Both Arendt and Fraser are asking us to pay close attention to our experience of origins and to do so in an unusual way. Usually we turn to origins for explanations. So, for example, we turn to the theory of evolution to explain where living things have come from, how they have changed, and how they might change in the future in a quest for understanding based on etiologies. More generally, we think that in finding the origin of some phenomenon, we have explained its existence. Arendt and Fraser challenge us to reverse our usual way of thinking about origins. Origins are not explanations but the source of our awareness of our temporal condition, our existence in time, and the connection between temporal existence and natality. We give birth to the new; we are the origin of creation in time.

Owning up to natality also is the source of our anxiety regarding the meaning of existence. To put this in terms akin to Fraser's, we exist without explanation, because with us explanation becomes a real issue for the first time. Questions like "Why do I exist?" or "Why am I here?" can turn us to questions like "Where did I come from?" for answers, but this last question takes us to origins. In other words, we circle back in our reasoning rather than getting to our goal: what raised the question of meaning in the first place was the experience of origin, the experience of birthing the new, the experience of natality, and our attempt to answer this question takes us back to its source.

Normally we approach these questions of meaning and existence as questions of finitude, that is, as questions that arise along with our awareness of our mortality. Montaigne was not the first philosopher to claim "that to philosophize is to learn how to die." Fraser's understanding of time's evolution is helpful in correcting this line of thinking as he emphasizes that temporal evolution is like the expanding universe: just as there is no space into which the universe expands and therefore the universe has no center or edge, so too time does not evolve into emptiness, that is, into eternity. The "expanding universe ... does not expand into a pre-existing empty space" but "it is space, with nothing external to it, which expands"; so too "time does not evolve within a preexisting expanse of time" (Fraser 48). If this is the case, there is no timeless void or abyss to threaten the meaningfulness of temporal existence.

Arendt moves us beyond the problem of death. She points out that "men, though they must die, are not born to die but in order to begin" (*The Human Condition* 246). All living organisms—including humans—die. But humans can do something else: we can begin.

Perhaps we focus on our finitude because death is easier to address than birth. Religions tend to provide a sense of the unreality of physical death even when they don't always promise an afterlife. Biology fits human existence into a longer and larger evolutionary schema wherein I have my genetic place: even though I die, my genes go on. I know my species will become extinct, but the time frame for this extinction stretches out so far I no longer feel personally threatened—it's not going to happen in my lifetime. And Western philosophy remains caught in the Platonic trap of debating whether or not there is an undying human nature, a timeless and eternal "soul" as the true "form" of the person. Indeed, philosophy even more so than theology or biology, seems responsible for putting us on this search for a "human nature" to begin with. But just as Fraser called on us to invert Plato's understanding of time and reality, Arendt refuses to speak of human nature, preferring to discuss the "human condition" as more adequate to our experience of existence: to be human is to be a "who," not a "what." And consistent with this, she asks us to think of our deaths in terms of immortality, or "endurance in time, deathless life on this earth and in this world" (*The Human Condition* 18),[3] rather than in the escapist terms of eternity.

The connection between origin and freedom made by both Fraser and Arendt is for the latter an explicitly political issue. If Fraser thinks in terms of biology, for Arendt the context for thinking about time is politics. In *The Human Condition* Arendt wrestles with what she sees as the loss of freedom that comes with the loss of politics. She begins by making the traditional philosophical distinction between the *vita contemplativa* or the contemplative life of philosophers and scientists, and the *vita activa*. She further divides the latter into labor, work, and action, and attempts

[3] On "human condition" versus "human nature," see 9–10 and 181.

to give these terms more specific meanings than is usual (7–8). Labor is human activity on earth oriented towards our physical survival and bodily existence; it is a biological process. Work is what we do to create the artificial or unnatural world in which we live. If the products of labor are meant to be consumed, the products of work are measured in terms of their use and durability.

Arendt's primary concern is delineating action as a first step to recovering a realm of human activity that she fears has been lost, that is, action as the activity that goes on between persons and depends on the human condition of plurality that is itself the condition for politics. She traces what she sees as the decline of politics that begins with Plato's concern in the *Republic* to use politics not for action but to enable philosophers to live contemplative lives. She asks us to step back behind Plato to an era when Greek political life was about freedom. The *polis* (state) used to be demarcated from the *oikos* (household): the latter was the place for taking care of the necessities for survival even as these extended into luxurious living (labor and work), but only in the *polis* could one be truly free to act, for in crossing over the boundary that separated house and state, one left behind necessity for freedom and full human existence. This place of freedom is where I live my individual existence, but it is only before others that I can become who I am as a human being capable of natality for I need witnesses to confirm the reality of my (free, unanticipated) actions.

Given this starting point, the cause for her alarm that today political life has been lost should be obvious. For example, we moderns often think of politics in terms of "political economy." But "economics" is etymologically derived from *oikos*, making the phrase "political economy" a contradiction when viewed from the ancient Greek perspective. When one entered the *polis*, it was by leaving the *oikos* behind. "Political economy" correctly describes current reality, however, insofar as political life has become a matter of administration, a household concern. Our political concern is more efficiency than it is novelty. This is why we turn to statistics to analyze human communities in terms of what we can predict about their behavior. But behavior is very different from action, and a concern with predictable behavior robs us of both the individuality and the freedom that are the results of our acting, our capacity to give birth to the new. Arendt expresses her greatest fear as follows: "The trouble with modern theories of behaviorism is not that they are wrong but that they could become true" which would mean "the deadliest, most sterile passivity history has ever known" (*The Human Condition* 322). We have become a consumer society, which is another way of saying action (and work) have been collapsed into labor whose products are consumed as necessary to survival—even when they're not. This society cannot be free.

Memory

The project of recovering politics and freedom might also be characterized as a recovery of memory. Who I am is disclosed in speech and action, according to Arendt, but action is "frail." It depends on others to witness it and perhaps more importantly, to remember it. *What* I am is a matter of information that can be stored in a data bank, but *who* I am depends on having the story of my words and deeds remembered by others. Arendt describes the origin of the *polis* in terms of the frailty that comes with this dependence: the *polis* was a place of "organized remembrance" (*The Human Condition* 198) wherein I was free to be who I was, where the new things I did and said would take on the permanence of being remembered.

Memory may thus be constitutive of my identity, but it also can be a trap. Are we limited to our pasts, and can we be anything more than what we have said and done before? Doesn't memory lead to loss of freedom and natality? Arendt raises this problem through describing the "irreversibility" of action. If what one says and does requires the presence of others as witnesses, that is, if the condition for action is plurality, once something is put out in public, it cannot be taken back. How, then, can we cope with the after-effects of action in such a manner as to preserve freedom? Arendt's common sense yet significant response is "forgiveness." Forgiveness has the capacity to carve out space for freedom within the legacy of the past entailed by memory. "Without being forgiven, released from the consequences of what we have done, our capacity to act would, as it were, be confined to one single deed from which we could never recover; we would remain the victim of its consequences forever" (*The Human Condition* 237). Forgiveness is the opposite of vengeance inasmuch as vengeance refuses to let go of the past and remains haunted by memory. But while forgiveness and vengeance are incompatible, there remains a possible connection between forgiveness and punishment: punishment can be an alternative to forgiveness; punishment, too, can liberate us from the past ("paying one's debt"), but only if punishment is understood as something other than vengeance. Arendt also admits that there are some cases of "radical evil" for which no adequate punishment can be devised.

In these cases, forgiveness also is impossible. (Arendt remains abstract in her discussion of radical evil and what might not be open to forgiveness, but one cannot help but think of Auschwitz and Arendt's reservations about the Nuremberg, Eichmann, and Frankfurt trials [*Eichmann in Jerusalem* 227–256].)

But in the normal course of action and our memories, forgiveness has the power to set us free. This is because granting forgiveness is itself an instance of natality. "Forgiving, in other words, is the only reaction which does not merely re-act but acts anew and unexpectedly, unconditioned by the act which provoked it and therefore freeing from its consequences both the one who forgives and the one who is forgiven" (*The Human Condition* 241). Where one might expect vengeance or some other reaction derived from the act, forgiveness responds by doing a new thing that ends the story of the action's after-effects and thereby liberates the persons involved and their community from the past. Memory need not be a trap.

What Arendt writes about forgiveness and natality should not cause us to lose sight of the dangers of forgetting and the liberating power of memory itself, however. Forgiveness sets us free from being condemned to live in the past with the after-effects of action. But remember, the context for action is plurality or political existence. In this context, memory itself can exercise a liberating power as it disrupts the official stories of the political communities in which we live and act. The history a community writes of itself too often aims at being a coherent master narrative serving an ideal sense of community identity. This ideological tendency is exacerbated by the experience that most histories are written from the perspective of the powerful in a community, that is, the stories are told by the "victors" in past struggles. In this way history actually can become an exercise in forgetting as actions that don't fit the master narrative are denied the status of history. What is needed in this case is a recovery of the memories that come "from below" and that constitute an alternative history of suffering or history from the "victims' " perspective.

These memories have been called "dangerous memories." This is how the German political theologian Johann Baptist Metz describes this kind of memory: it is "*a dangerous and liberating memory*" which "breaks through the magical spell cast by the ruling consciousness" and "lays claim to history as more than a screen against which we project our present interests"; it recognizes that "society's 'plausibility structures' can very much be 'obfuscation structures' "; it can see this because it "mobilizes tradition as a dangerous tradition and thus a liberating power" that "badgers the present and calls it into question"; and it has a future content based on the promise and hope contained in a tradition that compels us "to be in a continual state of transformation" (89).[4]

Memory when dangerous is not a trap; it rather is a liberating possibility. It is "dangerous" to master narratives as it recalls those actions and those actors whose stories we might forget because they won't fit into smooth narratives of progress or because their import extends beyond mere historicist concern with "facts." These dangerous memories liberate us from taking master narratives as the basis for assuming that the way things are is the way they have to be. They interrupt the present, remind us of our past hopes for a world different than the one in which we currently live, and goad us into the work of transformation in light of those hopes. And as we shall see below, their future content is in the form of a promise that makes the work of transformation reasonable. In short, dangerous memories afford the possibility that we can recognize the human temporal condition as situated between memory and hope.

Hope

Challenges to natality come not just from the past, however. Arendt recognizes the obvious problem with natality's future as well: if action flows from the principle of natality, if we are the origin of the radically new in reality, if we are free, then our actions are unpredictable. For Arendt, this unpredictability "is of a twofold nature": on the one hand, it is the "basic unreliability of men who can never guarantee today who they will be tomorrow"; and at the same time there is "the impossibility of foretelling the consequences of an act within a community of equals," the plurality that is the condition for action (*The Human Condition* 244). This is how action is different from behavior: behavior is in principle capable of being predicted by, for example, statistics. And this is why political philosophy rather than the social sciences provides access to understanding action, as was mentioned above. But if action is unpredictable, how can we ever come to understand it? Can we even think about the future if the future can be disassociated from the present and past, that is, if the

[4] See also Ostovich, "Epilogue: Dangerous Memories." Dangerous memories illustrate the way in which Metz takes seriously the disruptive power of eschatology and demonstrate why Metz is the originator of a "new" political theology; this differs from the "old" political theological concern with using theological categories as meta-temporal grounds for political values as seen in the work of Carl Schmitt (see Ostovich, "Carl Schmitt, Political Theology, and Eschatology").

future remains unpredictably opaque? Can we act reasonably if the future remains fundamentally up for grabs?

Arendt turns to the category of "promise" to deal with this challenge. A promise indicates a direction in which we will employ our capacity to give birth to the new, an orientation for our natality. Promises are like "islands of certainty in an ocean of uncertainty" (*The Human Condition* 244). They do not eliminate uncertainty; that is, they are not guarantees. But it is precisely the uncertain certainty of promises that respects human freedom and makes action possible. Promises are "the only alternative to a mastery which relies on domination of one's self and rule over others" (244). Any attempt to secure the future in the name of promises is misguided and doomed to failure: "when this faculty is misused to cover the whole ground of the future and to map out a path secured in all directions, [promises] lose their binding power and the whole enterprise becomes self-defeating" (244). Promises, like dangerous memories, point to the limits of our ability to master the future or the past. And these limits represent not a failure of power but a logical limit: we cannot reasonably be required to use our freedom to eliminate our capacity for natality that is the basis for our freedom without entering into a performative contradiction.

Arendt offers the biblical story of Abraham as an example of a life of promise. Understanding covenants to be formalized promises, Arendt credits Abraham with being the "discoverer" of "the power of stabilization inherent in the faculty of making promises." Abraham's story "shows such a passionate drive toward making covenants that it is as though he departed from his country for no other reason than to try out the power of mutual promise in the wilderness of the world" (243–244). Abraham is the model promise-maker. But he is something more, as Fraser points out: Abraham shows us what it is like to live under a promise or within a relationship based on promising. It is because he lives under a promise that Abraham is able to do the radically new and unexpected act—not just leaving Ur, but raising the knife over Isaac, his son and the embodiment of God's promise (*New American Bible*, Genesis 22).

Fraser approaches this story of Abraham by thinking about language. Fraser describes the "linguistic inertia" through which language as the bearer of cultural values resists change. But he also points out the paradox that language itself can be the source of changes in inertia. He quotes George Steiner's claim that "language is the main instrument in man's refusal to accept the world as it is" (167, qtd. in Fraser 27). This means "language ... has a capacity to change its direction of motion, for it offers ways through which the unpredictably new may be named" (Fraser 28).

This is what Fraser sees in Genesis 22: just as Abraham overcame the "phylogenetic inertia" that leads most living things to protect their offspring, he overcame the "linguistic inertia" evident in the fact that there was no "endearing phrase in ancient Hebrew to describe people who murder their children. Yet he was ready to do just that and offer Isaac for a cause that was hardly more than a name" (28). Abraham, a promise-maker living under a promise, could do the shockingly new.

For Fraser this is an example of "the freedom of the nootemporal umwelt" (28), but I am not convinced this is the appropriate temporal umwelt for understanding the story. Abraham's decision is so radical it introduces a rupture in the flow of nootemporality, a caesura that challenges us to think about not just Abraham but about time, ethics, and even thinking itself. Steiner is correct about language and the possibility of change, but there are some stories that introduce something so radically new they reduce us to silence. This is what happens to Søren Kierkegaard as he spends a lifetime trying to come to terms with Abraham and Genesis 22. He writes *Fear and Trembling*, but this writing is only possible for his "author," Johannes *de silentio*, "John *the silent*." And all he can do is ask: "no one was as great as Abraham; who is able to understand him?" (13). How are we to think through this story?

This question might also be phrased as, how we are to give an account of the hope that is Abraham's. Such hope characterizes a life of promise, but does this hope make any sense? Can hope be a reasonable way of living in the world? Is hope rational, and does it function as a critical principle? My concluding contention is that responding to these questions requires us to think further about the temporal level in which hope operates. Finding a reason for hope pushes us beyond the limits of nootemporality in the direction of the sociotemporal. Genesis 22, Abraham's binding of Isaac, disrupts the power of language to convey meaning. How can we understand the human temporal condition in light of such ruptures? We have here the opportunity to think through hope provided we take disruption seriously as a temporal category. And this is what eschatology is about.

Eschatology

In order to explain why I believe this is the case, I would like us to return to the context for action, for natality, for freedom, according to Arendt. This context is plurality exemplified by life in the *polis* or more simply, politics. And in this context let us ask a question of hope: what is it that we hope for? The classic answer to this question still has power: politics should reflect our hope for justice. Hope is what moves us to work for justice. But does this make any sense? Normally we think of ourselves as intentional animals: we use our minds to set goals and then figure out how to achieve them, and we are satisfied with this account of our "reason."

But if this is the case, how are we to understand our actions in pursuit of a just political order? We most likely have been acting for justice for as long as we have been aware of ourselves as members of groups, but where is the evidence that we have made the world more just by our actions? Certainly no one would claim we have achieved justice. In part this reflects the nature of an activity like justice. "Praxis" is the word Aristotle among others used for rational human activity in the political realm. This activity aims at the Good of which justice is an instance, but by its very nature this activity can never be complete: one must always do justice again. Still, shouldn't our actions make a difference in the world that lasts?

We also believe that there has been progress towards justice in some areas: the rights of women have been recognized even if those rights have not been achieved concretely; and we have ended slavery (at least manifestly and in some places) even if we have not overcome racism. We seem to be moving in the right direction. But what, then, are we to make of the long twentieth century with its development of total wars and technologically efficient modes of genocide? I do not want to get into cataloguing our failures in the pursuit of justice because these failures are both numerous and obvious. Justice is our goal in politics, but we have not achieved it and we have not made the world a more just place overall. At what point does the pursuit of justice become irrational? When do we say, "Enough, we've learned our lesson, we realize justice is impossible, now let's move on and do something else?" Clearly we're not going to say this, but can we explain why not? Is our hope illusory or even a delusion?

Among moral philosophers, Immanuel Kant may come closest to helping us work through the reasonableness of our hope for justice and the actions this hope informs. After all, Kant's practical reason frees us from failure by locating the moral worth of actions in ourselves and our capacity to do our duty regardless of the success or failure of our actions in realizing our intentions. We sparkle like gems in the independence of our reasoning and our rational power to do the right thing. But as J.B. Schneewind has pointed out, even for Kant, "the world must allow for the possibility of success." Moreover, "[i]f reason showed the perfect good to be a required but unattainable goal, reason would be at odds with itself"

(332–333). Unfortunately reason working on experience shows this to be the case with regard to justice.

And yet Schneewind's claim is suggestive: maybe the problem is with reality, specifically with what we assume to be the reality of time. We cannot deny injustice and the failure of our just actions to realize our goals. But the language of goals and rational ends depends on a particular understanding of time as linear and flowing in which the present floats between the past and future. We stand in the present and make our plans for the future taking into account what has worked or not worked in the past. This is the temporal reality of common sense, the time with which we are familiar—what Fraser calls "nootemporality." But as soon as we think about temporal existence in the context of justice, we should recognize how inadequate this understanding of time is: if time flows in one direction and the past is gone, what can be done about past victims of injustice? If the answer to this question is "nothing," I can only despair, for no amount of future justice can make adequate recompense to the dead. Even if we should find some way to make progress towards justice, no amount of progress can render the victims of injustice in history their due. We cannot think only about the future and still claim to be pursuing justice in a reasonable manner. Memory disrupts our way. It seems we are unable to explain why pursuing justice makes sense if we think about our experience and our temporal condition in linear terms, and linearity seems to be a key element of nootemporal thinking even given the complexity of nootemporality—remember Fraser's description of time's arrow.[5] Have we reached a dead end? Or are we now at a conflictual temporal boundary?

We have reached a place where we need to think some more and in other ways about time and about our temporal condition as situated between memory and hope. I would like to suggest that we have a potential resource for carrying on this thinking in the apocalyptic literature of the biblical era and the new concept of time communicated in the symbols and speculations therein: eschatology. This literature is found in passages in a number of biblical and extra-biblical books; it ranges in date of composition from the Book of Daniel in the Hebrew Scriptures to the Revelation to John in the New Testament; and most of it is found outside the biblical canon in books like Enoch. Wherever found, this literature is based on a revelation (apocalypse) by God to a (usually) pseudonymous

---

[5] Fraser 45–47, for example.

writer. These revelations address a variety of matters including speculation about the cosmic basis for calendars, the roles of angels, and explanations for the origin of evil in the world. Mostly, however, they address the question of justice. They typically are composed during periods of political oppression for orthodox believers and contain the explanation that current suffering is not punishment from God but the work of dark forces and the promise that this situation soon will end. The message is "hang on for the Kingdom of God is right around the corner," and this Kingdom will bring the realization of a just world order.

The temporal core of this literature is what interests me here. There is a way in which this literature encourages thinking about time in linear terms insofar as God's revelation usually includes "signs" of the end times and a schedule that helps the reader assess the current state of affairs (see, for example, Daniel 7–12 and the way in which these visions can be plotted in terms of historical events and personages). But a key element of these texts is thinking about time in terms of eschatology and the portrayal of the eschaton as an interruption of time's normal flow. The Gospel of Mark, for example, tells us to "learn the lesson of the fig tree," that is, learn to read the signs of the times, for the end is near; at the same time, however, Mark states, "But of that day or hour, no one knows," so "What I say to you, I say to all: 'Watch!'" (Mark 13: 28, 32, 37). Be prepared, and as Paul's letters and the Gospel of Matthew indicate, the person who is prepared already lives a certain way in pursuit of an ethical, just temporal order (Matthew 24–25; 1 Corinthians 15:50–58; 1 Thessalonians 4:13–5:22). Furthermore, it makes sense to live this way: the Kingdom of God is about political justice; your hope for that kingdom means it makes sense to work for the promised justice now, and this in full awareness that your effort is not what brings about the kingdom but is the rational response to God's promise of justice.

It might be objected that a two-thousand-year-old tradition of apocalyptic literature hardly is the place to turn for responses to contemporary conflicts. If the eschatological understanding of time as disruption is helpful in coming to understand the reasonableness of pursuing justice, why has no one made use of this understanding before? I could respond in terms of the ongoing lives of classic texts in which new meanings become available only in light of new problems, but something else is going on here. The problem is rather that we have never taken seriously the temporal rupture described in these texts. Instead the tendency has been to escape their temporal tensions by collapsing them into a reading based on linear time wherein the "end" is a stage in time's flow from the past through the present to the future. This linear understanding of time informs the way in which the political message of apocalyptic texts often has been privatized: the hope for justice has been translated theologically to a concern with my receiving justice at the end of my life in the afterlife; the day of judgment has become the day of my death; and the Kingdom of God has been misread as a spiritual phenomenon, that is, heaven, and removed from the realm of politics.

Or eschatology has been reduced to teleology in the philosophy of history, as described above. The result is an unintentional irony: eschatology has been secularized as progress in an attempt to "tame" the political excesses of apocalyptic movements, but the result is to twist eschatological hope into the utopianism that truly is dangerous. Grafting eschatological hope onto a time line has turned that hope into the goal of a historical process that serves as rationale for acting "as if" that goal were present already. It becomes easy to sacrifice the present—and the people who live in the present—to a utopian vision of the future that will come true if we work hard enough and make enough sacrifices (including sacrificing others) to achieve it.

It might also be objected that I have smuggled in a guarantee of justice as the ground for the meaningfulness of just actions, a guarantee which takes the form of a God who is the source of the eschatological revelation and who is omnipotent, omniscient and eternal. This is the God of Christian theism, and belief in this God could be seen as turning us from trust in a promise to confidence based on a guarantee. After all, by definition this theistic God is in complete control and cannot fail—that's why we're tempted to believe in God. If this is so, we are again in the self-defeating position Arendt identifies with those who want to map out the future completely in the name of freedom. And we would remain trapped in a linear temporal umwelt.

From a religious perspective, this need not be the case. There are biblical alternatives to the God of theism especially in the Hebrew Scriptures. For example, our English translations of Exodus 3:14 usually work from the Greek translation of the Hebrew Bible (the Septuagint) in reading this verse as "I am who am" and the absolute "I AM sent me to you." This makes sense from a Greek perspective in which God serves as the metaphysical ground for reality and time because God is eternal or timeless. The passage in Hebrew reads very differently, however, playing on the "name" of God as YHWH, a word derived from "becoming" and that testifies to a God who is dynamic, not an unchanging ground, and who is intimately present in time and history, not eternal. Reading apocalyptic literature against the background of this Hebraic God returns us to the realm of working for justice as a rational activity.

I am more interested, however, in asking what apocalyptic eschatology would look like without God. This is not to take a stand on whether or not God exists; it is much more a matter of taking eschatology seriously as a way of thinking about time and reason rather than as a religious perspective. Thinking eschatology through without God becomes the equivalent of thinking through reason without metaphysical foundations. Paying attention to origins and our capacity to be originators involves a shift in our thinking. For example, we have already seen Fraser refer to Heidegger's concept of "thrownness" (*Geworfenheit*) above: we are cast into a world not of our own making. But the concept of natality seems to entail turning this around: we are the originators of the world in which we exist and in which the feeling Heidegger describes is a result of sharing that world with other freely acting originators. This is the concept of origin (*Ursprung*) one finds in the work of Heidegger's contemporary Walter Benjamin ("Epistemo-Critical Prologue." *The Origin of German Tragic Drama* 27–56).

Benjamin, a secularized Jew, also helps us open a window onto a life in eschatological hope and how this hope helps make sense of pursuing justice. At the end of "On the Concept of History," Benjamin writes, "We know that the Jews were prohibited from inquiring into the future: the Torah and the prayers instructed them in remembrance. This disenchanted the future, which holds sway over all those who turn to soothsayers for enlightenment." These last are our planners and our utopian visionaries who look to realize justice in the short- or long-term future. Benjamin continues, "This does not imply, however, that for the Jews the future became homogeneous, empty time. *For every second was the small gateway in time through which the Messiah might enter*" (*On the Concept of History* 397, emphasis added). The messiah, the Davidic king, the son who is to rule in God's kingdom, is the bringer of justice. Our hope in his arrival does not mean rationalizing our seeking justice in terms of some future utopia and justifying current actions in terms of this goal; we are not to live "as if" the kingdom were already here. It is rather a matter of acting with the hope that any moment the flow of time might be disrupted in the eschaton. If this is our hope, we act "as if not," that is, as if the world were not condemned to injustice, as if this world were not a place where the pursuit of justice made no sense. And this hope, a hope in which everyone can share, makes sense of our just acting.

This never will be easy. Hope means living with what Fraser describes as "permanent revolution." And Fraser issues a further, tougher challenge for us: If change comes through conflict, we must figure out "how to perpetuate conflicts" (267). There is a tragic quality to life. But the rational response to this tragic quality is not resignation. It is hope.

References Cited

Arendt, Hannah. "Auschwitz on Trial." *Responsibility and Judgment*. Ed. Jerome Kohn. New York: Schocken Books, 2003. 227–256.

——. *Eichmann in Jerusalem: A Report on the Banality of Evil*. New York: Penguin Books, 2006.

——. *The Human Condition*. 2nd ed. Chicago: University of Chicago Press, 1958.

Aultman, Mark. "Nootemporality, Sociotemporality, and the Internet." *KronoScope: Journal for the Study of Time* 9.1–2 (2009): 129–145.

Benjamin, Walter. "On the Concept of History." *Walter Benjamin, Selected Writings, Volume 4: 1938–1940*. Trans. Harry Zohn. Ed. Howard Eiland and Michael W. Jennings. Cambridge, MA/London: Belknap Press of Harvard University Press, 2003. 389–400.

——. *The Origin of German Tragic Drama*. Trans. John Osborne. London/New York: Verso, 1985.

Fraser, J.T. *Time and Time Again: Reports from a Boundary of the Universe*. Leiden/Boston: Brill, 2007.

Huxley, Julian. *Religion without Revelation*. New York: Mentor Book, 1957.

Kierkegaard, Søren. *Fear and Trembling*. Trans. Alastair Hanney. New York: Penguin Books, 1985.

Löwith, Karl. *Meaning in History: The Theological Implications of the Philosophy of History*. Chicago: University of Chicago Press, 1949.

Mayr, Albert. "The Round Table: A Social Time Sculpture." *KronoScope: Journal for the Study of Time* 9.1–2 (2009): 111–119.

Metz, Johann Baptist. *Faith in History and Society: Toward a Practical Fundamental Theology*. Trans. J. Matthew Ashley. New York: Herder and Herder, 2007.

Montaigne. "That to Philosophize Is to Learn How to Die." *Selected Essays of Montaigne*. Trans. John Florio. Ed. Lester G. Crocker. New York: Pocket Books, 1959. 16–35.

Needham, Joseph. *Time: The Refreshing River: Essays and Addresses 1932–1942*. London: George Allen and Unwin, 1943.

Ostovich, Steven. "Carl Schmitt, Political Theology, and Eschatology." *KronoScope: Journal for the Study of Time* 7.1 (2007): 49–66.

——. "Epilogue: Dangerous Memories." *The Work of Memory: New Directions in the Study of German Society and Culture*. Ed. Alon Confino and Peter Fritzsche. Urbana/Chicago: University of Illinois Press, 2002. 239–256.

Schneewind, J.B. "Autonomy, Obligation, and Virtue: An Overview of Kant's Moral Philosophy." *The Cambridge Companion to Kant*. Ed. Paul Guyer. Cambridge: Cambridge University Press, 1992. 309–341.

Steiner, George. *After Babel: Aspects of Language and Translation*. Oxford: Oxford University Press, 1975.

BACHELARD'S "DISCONTINUOUS BERGSONISM" IN ZORA NEALE HURSTON'S "SWEAT": HOW SELF-GENERATION OF "PURE TIME" ENGENDERS FREE CHOICE

Patricia McCloskey Engle

Abstract: Zora Neale Hurston's short story "Sweat" presents a practicable model for generating the experience of "duration" or "pure time," as conceived by Henri Bergson and further developed by Gaston Bachelard. The protagonist of the story demonstrates not only the nature and benefits of pure-time experience, but also the disposition and behaviors necessary to its initiation. Employing literary criticism, application of Bergson's and Bachelard's duration-specific theoretical premises, and biographical grounding on Hurston's leanings with respect to time-consciousness, this article probes evolutionary implications for self-generated shifts in temporal experience, initially in willful invocation of passive resistance to re-direct aggression back upon its originator and, in a larger context, in the power that people may gather when they demand their right to make their *own* time and hence, their *own* decisions.

Keywords: Duration, pure time, Bergsonism, Bachelard, Hurston, female temporality, detachment, causality, linear time, slavery, domestic abuse, Jung.

Literary fiction, as we know, can provide rich inspiration for interdisciplinary inquiry into the nature and limits of human time-experience. Kurt Vonnegut's imaginative representation of how time is experienced by creatures on the fictional planet Tralfamadore, for example, in his epic, 1969 novel *Slaughterhouse-Five*, inspired physicist Fred Alan Wolf's development of the "new physics of time" unfolded in his 1981 *Taking the Quantum Leap* (Rackstraw 49–50). Zora Neale Hurston's short story "Sweat" may not have directly inspired Gaston Bachelard's conception of "discontinuous Bergsonism," which was outlined in his 1950 *Dialectic of Duration* (29) and will herein be explained, but it provides a useful case study for practical application of his theory and the benefits to be derived from experiences of what Henri Bergson termed "la durée," also known as "duration" or "pure time." My study of the story's protagonist, Delia Jones, is intended to add to Bachelard's duration model by explicating the "front end" work needed to generate duration, specifically in terms of attitude and behaviors, some of which are more clearly seen through the lens of literary criticism.

In this 1926 tale of a battered wife who sees her husband's cruelty against her play out to his own undoing, Hurston gives us a protagonist who engineers, with impeccable timing, those strategic breaks and rhythms in time that Bachelard contended *must* be self-generated, in order for people to create for themselves the only time-experience capable of permitting performance of a "free act" by a "definite person" (Bergson *Time and Free Will* 165). Delia's improvisational use of "street theater" drama to break the momentum of her husband's will constitutes "behavior initiating duration" by Bachelard's standards (59). As well, the transformative, pre-sleep contemplation in which Delia engages on two particular nights detailed in the story demonstrates experience of not only the "hors-temps" or "time-outside-time" that Irma Garcia sees as unique to "femalear" temporality (177), but in fact the "pure time" that Bergson and Bachelard saw as essential to autonomous selfhood for all peoples. Delia could not, however, find her instinctual way to behavior instigating duration if she did not work hard at maintaining a disposition favorable to duration, and here is where the story's title yields useful symbolism. In order to avert the ignominious future toward which her downtrodden origins and abusive husband drive her, Delia must cultivate and maintain an operational base of detachment and "agape," or love without self-interest, and she must balance her life in complementary rhythms of economic self-sufficiency and extemporaneous creativity. In order to maintain such purity of intention, she must fastidiously "cleanse" herself of her husband's toxic aggression. The implications with respect to this volume's theme of "origins and futures" are compelling, demonstrating the new and better futures that people can create for themselves by re-directing concerted forces of social oppression back upon their originators, as happens to the abusive husband in "Sweat." Before unfolding these parallels between story and theory, I offer a brief overview of those points of convergence and divergence between Bachelard and Bergson most immediately relevant to this discussion, a sampling of Hurston's stated "time" interests that suggest her conscious attempt to give us something time-revolutionary in this story, and the plot line of "Sweat."

Bachelard took no exception to Bergson's initial conception of duration as time separate from space and permitting experience of simultaneity, at least with respect to the past and present. Specifically, Bergson proposed

> ... two possible conceptions of time, the one free from all alloy, the other surreptitiously bringing in the idea of space. Pure duration is the form which the succession of our conscious states assumes when our ego lets itself live, when it refrains from separating its present state from its former states. (100)

Neither did Bachelard dispute Bergson's idea that the purpose of separating time from space is to free the human will from "mechanistic causality" ("Henri Bergson"). While Bergson in *Time and Free Will* saw the linear impetus of the Western conception of time creating "a tendency to transform relations of succession into relations of inherence," thereby working against experience of duration (208), Bachelard likewise expressed concern that "causality and form dominate time and space" (73). However, while Bergson saw pure time as a natural phenomenon accessible from both "within" and "outside" of us (*Time and Free Will* 226–227), Bachelard clarified that "[f]ar from innermost duration being a property we own, it is a work we create, and is always preceded by an action centered on an instant" (55). Moreover, because Bachelard saw "the decisive centers of time" to be "its discontinuities" (54), he took exception to Bergson's conception of duration as a single, continuous flow of time; Bachelard found it "impossible not to recognize the need to base complex life on a plurality of durations that have neither the same rhythm nor the same solidity in their sequence, nor the same power of continuity" (19). As Conrad Russell explains, Bachelard's problem with Bergson's duration is that it gives "rise to a 'viscosity', which binds past to present with the force of an efficient cause" (15). Thus, Bergson's descriptions of pure time did not, in Bachelard's view, effectively purge causality's sustained play on the will. In all other regards critical to my application of Bachelard's theory in this paper, he agreed with Bergson, particularly sharing Bergson's central concern that without access to pure time, our lives "unfol[d] in space rather than in time; we live for the external world rather than for ourselves; we speak rather than think; we 'are acted' rather than act ourselves" (*Time and Free Will* 231). Clearly, both philosophers shared a common goal of raising the human condition toward higher possibilities through a more productive experience of time—an interest, I suggest, that Hurston also shared.

Hurston was a folklorist and writer, remembered especially as the female voice of the 1920s Harlem Renaissance. Her falling out with leaders of that movement is said to have stemmed from two things, primarily: her insistence on keeping character dialogue true to rural, Black southern dialect and her unwillingness to directly confront the "race problem" in her work. To Hurston, black lives were "not reactionary. Black folks weren't living in reaction to what whites had done to them" (Fields 21), and indeed, this is how the trouser-wearing, pistol-toting Hurston seems to have lived her own life. Her insistence upon a "non-reactive" life for *all* people, as will be discussed, is the central theme in Hurston's work that aligns her with Bachelard's more precise call for suspension of causality, so "that continuous duration has its chief function removed" (60). Hurston took her B.A. in anthropology from Barnard University in 1928, the year of Bachelard's first publication and two years after she had published "Sweat." She compiled folklore collections during field work in the Caribbean and Honduras, as well as the deep rural south of the United States, under the direction of Franz Boas, and trained with hoodoo practitioners, undergoing initiation ceremonies that included three days of a fasting vision quest, lying naked upon a magical snakeskin cover (Hurston *Mules* 188). Though she left behind an impressive body of work spanning multiple genres, Hurston died penniless, forgotten, and largely out of print, in a Florida welfare home in 1960, to be rediscovered by a new generation of writers in the mid-1970s.

Hurston's memoirs document her bracingly inquisitive relationship with time. Of note is her imaginative conception of time as separate from and disdainful of the material world, which she distinguishes from "Time" as "Thing":

> Thing lies forever in her birthing-bed and glories. But hungry Time squats beside her couch and waits. His frame was made out of emptiness, and his mouth set wide for prey. Mystery is his oldest son, and power is his portion.
>
> That brings me before the unlived hour, that first mystery of the Universe with its unknown face and reflecting back. For it was said on the day of first sayings that Time should speak backward over his shoulder, and none should see his face, so scornful is he of the creatures of Thing. (*Dust Tracks* 322)

This positing of time in contravention to material desires demonstrates Hurston's understanding of the doomed nature of any embrace of time as the vehicle of conveyance to future material gain—or perhaps, *any* gain or reward, even as anticipated in the religious sense of the Judeo-Christian tradition, which originated the Western mind's conception of time as linear, future-directed impetus (Fraser 22). Likewise, Hurston's "Time vs. Thing" conception suggests that glimpses of the future or "future-vision," as such experiences shall herein be termed, must necessarily be denied to anyone whose life is governed by acquisition of Thing's endless progeny. Hurston's horrific representation of Time's delight in devouring such ambitions asks what, if any, alternative is possible. In both her life and her work, she seems to have envisioned not only an answer to that question, but also a process for realizing the vision that aligns to Bachelard's "discontinuous Bergsonism" with uncanny precision.

Because she understood time's "beneficent moods," Hurston in *Dust Tracks on a Road* claimed to have achieved temporal perspectives that sound like both deep time and future-vision:

> He [Time] has commanded some servant moments to transport me to high towers of elevation so that I might look out on the breadth of things. This is a privilege granted to a servant of many hours but a master of few, from the master of a trillion billion hours and a servant of none. (322)

From perspectives thus achieved, she discovered that one requisite to liberation from Thing's control is an abiding respect for "destruction and construction [as]... but two faces of Dame Nature" (*Dust Tracks* 348). An early indication of Hurston's understanding of the value of such complementary rhythms is her admiration, as a child, for the Old Testament David, deftly moving between destructive and creative acts: "he smote 'em hip and thigh. Then he sung songs to his harp awhile and went out and smote some more" (*Dust Tracks* 54). Hurston frames her own life in these same terms of simultaneity in time and complementarity in deed: "I have stood on the peaky mountains wrappen in rainbows, with a harp and a sword in my hands" (280). Knowing that she could not "pry aloose the clutching hand of Time,' she vowed to "turn all [her] thoughts and energies on the present" (284). In remarks like these, Hurston emerges as a balancer and keeper of time, and more—as a co-operative generator of duration in all the most practicable and evolutionary terms that Bergson and Bachelard envisioned. In effect, the notion of duration conceived by Bergson and modified by Bachelard raises the high-water mark in the wells of human integrity by challenging people to respond to oppression with actions untainted by oppression's degrading effects. It is in this latter regard that the example of Delia Jones in "Sweat" works most powerfully, as a model of distinctively non-reactionary discontinuous Bergsonism.

"Sweat" takes place over the course of one full summer, starting "at eleven o'clock of a Spring night" (Hurston 136) and ending in the "[d]og days" of August (142). As the story opens, protagonist Delia Jones has been married to Sykes Jones for fifteen years (137); Delia recalls that Sykes started beating her two months after their wedding (138). "[P]ritty ez a speckled pup" in her youth (140), Delia is now thin, with "stooped shoulders" (137) but "muscled limbs" (138). She is self-employed as a laundress, having used her earnings to buy the home she shares with Sykes and everything in it, as well as the grounds that she lovingly planted (138) and the horse and wagon they share (136). She is a spiritual woman of "habitual meekness" (137) who consoles herself singing gospel songs. Sykes is a self-centered, hedonistic brute. He is or was employed, possibly as a chauffeur (140), but spends his wages on his own pleasures, outside their home. "Sweat" is about Sykes's attempt to remove Delia from their home so that he can install Bertha, his latest mistress, to replace her. When Delia realizes his intentions, she surprises him by defending herself for the first time, raising a skillet against him (138). Sykes then resolves to drive her away by preying on her fear of snakes, staging a rattlesnake in a box outside their kitchen door. When that plan fails, he places the hungry diamond-back in Delia's lidded laundry hamper, to attack and presumably kill her when she returns home one Sunday night (145). At the sight of the snake she flees to the hayloft of her barn, awakening next morning to hear Sykes breaking up the snake box. She watches him enter the house after listening at the door long enough to confirm, by the quiet within, his expectation that Delia is dead and the snake's venom spent. She hears, as she creeps toward the house, first the "ventriloquist" whirr of the snake's threatening rattle and then the commotion as Sykes and the snake battle to the death (146). Knowing that they are too far from a hospital to save Sykes, Delia feels a "surge of pity," watching him crawl to the door, calling her name, his neck "horribly swollen … and his one open eye shining with hope." She turns away, knowing that in these next minutes, "the cold river was creeping up and up to extinguish that eye which must know by now that she knew" (147).

The character exposition in this story gives us nothing of the couple's backgrounds prior to their courting, but we can expect that their respective ancestral lineages were bound together in the US slavery experience. In this respect at least, Delia and Sykes share the same origins. Sykes embodies and recreates slavery patterns. Seeing that young Delia had attracted the interest of other suitors, he resolved to win her. Though he had to eat "some mighty big hunks uh humble pie" to persuade her to marry him (140), his goal all along had clearly been to own, exploit, and dispose of Delia. To external appearances, he may seem to have succeeded. As Joe Clarke, on whose store-porch the village men gather, observes, Sykes is one of those men

> … dat takes a wife lak dey do a joint uh sugar-cane. It's round, juicy, an' sweet when dey gits it. But dey squeeze an' grind, squeeze an' grind an' wring tell dey wring every drop uh pleasure dat's in 'em out. When dey's satisfied dat dey is wrung dry, dey treats 'em jes' lak dey do a cane-chew. Dey thows 'em away. Dey knows whut dey is doin' while dey is at it, an' hates theirselves fuh it but they keeps on hanging' after huh tell she's empty. Den dey hates huh fuh bein' a cane-chew an' in de way. (140)

For Sykes, like most abusers, the past and future are one continuous, unbroken flow of repeated aggression, first directed against his forebears, through the institution of slavery, and then delivered by his own hands, against his wife. Despite the opportunities that marriage presents him, to deflect and moderate his aggressions, Sykes never breaks from the slavery conventions he inherited. He remains a slave to his corporeal desires and a slave owner in his attitude toward his wife. Delia, on the other hand, seems to have lived, on some level at least, purposefully framing her future—and Sykes's too, if he could share her vision of it—in outcomes distinctly different from that which their shared origins seem to dictate.

The "porch sitters," as Myles Raymond Hurd refers to the village men (12), also tell us that the rhythm to Delia's laundry service is as regular as clockwork. Watching her pass in her cart, Joe Lindsay comments, "Hot or col, rain or shine, jes 'ez reg'lar ez de weeks rool roun Delia carries 'em an' fetches 'em on Sat'day" (139). On Sundays, she attends church services, staying for the evening "love feast" (144); drives home; sorts laundry, "put[s] the white things to soak" (136) and goes to bed. We are given to understand that laundry scrubbing, drying, and folding, along with cooking and gardening, take up the rest of her week, and the routine begins again. As will be discussed, maintenance of a weekly routine that balances her paid labor with extemporaneous expressions of creativity provided by her gardening and singing satisfies at least threshold measures of Bachelard's prescription for disposing oneself to duration, since "the real action of time requires the richness of coincidence and the syntony of rhythmic efforts" (20).

One aspect of Delia's work routine bears closer consideration: Delia's Sunday-night labors cause Sykes to castigate her as a religious hypocrite. Her defense that she must start on Sunday night in order to "git through by Sat'day" (137) shows that Delia is no blind follower of dogma; she thinks for herself, and for this reason alone, interpretations of "Sweat" as a story of a spiritual woman's simple trust in divine retribution must fail. Rejection of the "blind faith" interpretation of Delia's longstanding passivity is also supported by Hurston's own declarations with respect to the function of religion:

> People need religion because the great masses fear life and its consequences.... Feeling a weakness in the face of great forces, men seek an alliance with omnipotence to bolster up their feelings of weakness, even though the omnipotence they rely upon is a creature of their own minds. (*Dust Tracks* 277–278)

With respect to the place of religion in her own life, Hurston acknowledged a responsibility to use mind and will that aligns her even more strongly with Bachelard's vision of self-generated duration: "I do not pray. I accept the means at my disposal for working out my destiny. It seems to me that I have been given a mind and will-power for that very purpose" (*Dust Tracks* 278). Hurston imbues Delia with a similarly strong strain of self-reliance. Attendance at weekly services is the *only* venue for her social interaction that the story gives us, and she conducts it on her own terms. The "love feast" service that she enjoys on Sunday evenings is a Baptist celebration of Christian agape; as Hurston describes in this footnote to her memoirs, "[s]ince no one is supposed to take Communion [at the love feast] unless he or she is in harmony with all the other members, there are great protestations of love and friendship" (*Dust Tracks* 266). To avail herself of the love feast's full benefits, Delia transfers her church membership to the neighboring town of Woodbridge "so Ah don't haftuh take no sacrament wid [Sykes]" (143). The hymn that she sings, with the refrain "Jurden water, black an' col / Chills de body, not de soul / an' ah wantah cross Jurden in uh calm time" (144) sustains her, presumably extending her experience of duration on the long ride home and throughout her solitary toil, while also serving as a mantra to invoke duration.

As previously noted, Delia's work-week rhythm is also significant because it represents a ritual of complementarity maintenance, as she balances the "masculine" spheres of economic self-sufficiency, property ownership, and at least the effective appearances of self-defense, with "feminine" pursuits of artistic beauty—her singing and gardening. Her "little home" is entirely of her extemporaneously inspired creation and born of future-vision: "She had built it for her old days, and planted one by one the trees and flowers there. It was lovely to her, lovely" (138). Katherine Lee Seedel's interpretation of Delia's gardening and singing "as a metaphor for the work of the human creator, that is, the artist" (177) is relevant here, in that it highlights Delia's role as the agent of complementarity in a marriage of opposites. As Wilfred D. Samuels observes, Hurston's fiction frequently seeks to "locate a male-female relationship whose success is grounded in meaningful reciprocity" (240). In this regard, Sykes's refusal to check his aggression through adaptation or even accommodation of any of Delia's "feminine ways" makes his a story of failed maturation—for our purposes, confinement to the poor future dictated by poor circumstances of origin—but his perverse obstruction of his own development in no manner curbs or perverts Delia's growth. Even after Sykes brings the rattlesnake to drive her from her home, she continues to cook for him and advances "timid friendliness" toward him (142). Even on the night he tries to kill her, Delia's first impulse, when she comes home and finds the snake-box empty, is to hope that Sykes has undergone a change of heart and gotten rid of the snake for her (144). In these open-hearted overtures, Delia is neither gullible nor beaten; she is something much nobler and necessary to the generation of duration. When the narrator cites Sykes's repulsion of Delia's patient overtures as evidence that "the breaches [between them] must remain agape" (142), Hurston's pun on the "higher love" of Christian "agape" becomes apparent. As Hurd notes,

> As an adjective, it [the term "agape"] heightens our awareness of the seemingly unbridgeable connubial disharmony between the incompatible couple. On another level it signals a shift from sexual to nonphysical terms of correspondence in their marriage, and from a third angle of vision, it refers to love feasts mentioned in both Jude 12 and the story itself. (11)

This coded reference to Christian agape in the story brings to light the critical function this transcendent, "higher" love plays in the generation of duration, since Bergson himself considered "love ... to be the very essence of the creative effort" (*Two Sources* 95).

Sykes, after all, wants nothing more than to unseat Delia from who she is and make her like him: desperately enslaved to greed, fear, and hate; unable to claim or even imagine her own time, creative potential, destiny, or personhood. As such, Sykes represents the meaning of "real slavery," as Hurston understood it. According to Hurston, the institution of slavery in the US "was only an outside symbol. Real slavery is couched in the desire and the efforts of any man or community to live and advance their interests at the expense of the lives of others" (*Dust Tracks* 283). In a sense, Delia's home is who she is; she is not the body that Sykes has exploited, and her home represents her time. If she is to retain her selfhood, though, she must generate an experience of time that is unbounded and free of the mechanistic causality induced by spatialized time. If she does not respond to Sykes's manipulations with willful invocation of pure time, she will be limited, as Bachelard asserts, to committing only the "reflex act stuck in the instant and taking its whole meaning from some spatio-temporal coincidence" (57)—and she will lose all. Sykes's actions against Delia are calculated to meet his selfish designs, but nothing he does can define his wife's life or time. Primarily through demonstrations of behavior uncharacteristic of her but pointedly understandable to Sykes, Delia creates "breaks" in the temporal momentum Sykes seeks to exert over her.

The first such break is Delia's aforementioned decision to quit the local church she had attended with Sykes and join a neighboring congregation. That self-invoked removal from place, repeated each Sunday, mirrors Delia's self-invoked removals from time and emphasizes the ceremonial nature of them. As Bachelard suggests, "one can never over-emphasize the importance of this consecration of beginnings" (57). Similarly, the "new Delia's" demonstrations of self-defense, such as the skillet-brandishing scene (138) and her shocking "Ah hates you tuh de same degree dat Ah useter love yuh" tirade (143), are strategic improvisations designed to free those moments from the momentum of Sykes's will and frame at least the bridge to Delia's pure-time forays, to be pursued at her later convenience. It is significant, as Hurd notes, that Delia "restrain[s] herself [in these scenes] from taking any physically assertive, compensatory steps to end her unhappiness" (10). Her demonstrations of self-defense are for appearances only and intended to stun Sykes, which they clearly do.

By disarming her husband in this way, Delia invokes the power of pause to assert her own "willed order" to the situation (Bachelard 89), without dropping to Sykes's level of degradation. As Bachelard explains, such "deferring [of] an action means that its causality is suspended.... No longer does the stream impel the stream. We are free to decide what is urgent" (60). To feed into Sykes's aggressions would be to surrender her "original" and wholesomely evolving self; so Delia essentially "stops time"—ordinary time, that is, or the "journées" of female time-experience—to force an "instant" and open "hors-temps," which, according to Garcia, is "situated totally on the fringes of time, where silence treads," and corresponds to "moments of recuperation in the narrative[s]" of female writers (177). Garcia further contends that "[t]he delight that woman experiences in insisting on these hors-temps which she can manipulate at will, is perfectly understandable; she alone can make decisions about how long each of them will last, [and] she can assign to them her own scenario" (177). Again, though Garcia's temporal categories are intended to describe only female time-experience, initiation of these states as process steps corresponds very closely to duration-generation strategies suggested by Bachelard. Indeed, if time is (as Hurston and as Lewis Carroll's "Alice" suggest) a "he," then "he" may well require a "femalear" response to unlock and mobilize his higher possibilities.

One could argue that Delia forges rhythm disruptions simply to "buy time" to think—to decide how to react in order to "best" her spouse. Russell rightly sees Bachelard's concept of affirmative, discontinuous Bergsonism as animated by negation: "If you expend energy in a decision or affirmation it is because you are overcoming an obstruction, created by someone or something. All conscious thought involves undermining, negating, appearances. The real is analyzed 'under the blows of negation' " (5). The key here is the pre-eminence of the "non-reactionary life," as Hurston intended, which aligns precisely with the model that Bachelard envisioned. In Delia's removals to pure time, she is clearly doing more than thinking; she is purging Sykes's corruptive influences and recalibrating her original self. Only after completing that process, in pure time, is she able to initiate, as Bergson suggests, the "free act" of a "definite person" (*Time and Free Will* 165) as opposed to the knee-jerk response of a "conscious automaton" (168). Likewise, one might cite Delia's having "built [her little home] for her old days, and planted one by one the trees and flowers there" (138) as evidence of ambitions for her own future gains, leaving her as vulnerable to Time's appetite for "Thing" as Sykes, for his schemes to dispossess Delia from her home and plant Bertha there in her stead. Aside from the critical difference posed by Delia's having *worked* for all she owned, her sharing all she owned with Sykes for fifteen years—room, board, food, and transportation, at least—certainly serves to offset the momentum of aggressive acquisition, as does the creative energy she expends toward gardening and singing, for sheer love of beauty.

Hurston, no stranger to the stage, well understood the function of rhythm disruptions as a tool of creative expression and personal growth. Noting the difficulty with which "white dancers" negotiate the "abrupt and unexpected changes" in African American music, she said:

> There is always rhythm, but it is the rhythm of segments. Each unit has a rhythm of its own, but when the whole is assembled it is lacking in symmetry. But easily workable to a Negro who is accustomed to the break in going from one part to the other, so that he adjusts himself to the new tempo. ("Characteristics" 835)

The breaks Delia engineers through her demonstrations of dramatic improvisation call to mind so many avant-garde, countercultural art movements that ended up lulling the populace into consumerist or narcotic escape. What Hurston and Bachelard propose is not escape, however, but engagement and, in fact, activism of the highest and purest degree—the right of all people to a well-lived life, a *pura vida*, as the Costa Ricans say.

Likewise, Delia's "calm" declaration to Sykes that "Ah hates you ... tuh de same degree dat Ah uster love yuh" (143) persuades critics such as Robert Bone to see the story as a "revenge fantasy," involving a "man and wife ... locked in a mutual hatred" (qtd. in Hurd 8). Delia is no saint; she is certainly capable of anger. Consider her reaction as she observes the snake's "chalky-white fangs" clamp through the wire mesh of its cage: "She stood for a long time in the doorway in a red fury that grew bloodier for every second that she regarded the creature that was her torment" (143). She does not act under the influence of such anger, though; she engages in that fourth category of "femalear" time, under Garcia's system, termed "attentes," or "times of waiting" (177). Again, choosing this course of non-action neatly fits Bachelard's model; "[p]ure consciousness," he predicted, "will be revealed as the capacity for waiting and for watchfulness, as the freedom and the will to do nothing" (18). Delia's careful wording of her declaration of "hate" warrants scrutiny; she describes a hate as indifferent and devoid of self-interest as her love for Sykes had been. She had loved Sykes without need to control him, and when she hates him, it is without need to destroy him. Sykes's love and hate work in quite the opposite fashion. Again, there is more going on here than Delia's simply stepping out of the way to allow Sykes to reap what he has sown. In the process of generating duration, Delia purges herself to ensure that no seed of his has taken root in her. The couple's childlessness, like Delia's occupation as laundress, may operate on a symbolic level in this regard. This notion that duration generation presumes inclusion of self-purification in the initiation stage is consistent with the principles of the non-reactive, self-directed life at the heart of Hurston's and Bachelard's work. Examination of Delia's most prominent pure-time experiences demonstrates the importance of this step.

Hurston's descriptions of Delia's "interior" times are scant and sometimes second-hand, but they allow enough to support that Delia makes her decisions during episodes of pure time. The Sunday night "love feasts," as previously described, are likely communal invocations of pure time, but we do not see her making decisions or choices while at church. The two instances of pre-sleep contemplation she induces provide stronger demonstrations. In the first, as she lies in bed the night of the skillet-brandishing scene, she reviews "the debris that cluttered their matrimonial trail," noting that "[a]nything like flowers had long ago been drowned in the salty stream that had been pressed from her heart. Her tears, her sweat, her blood" (138). She holds to the future-vision of her "lovely" home and resolves to do nothing. Achievement of this resolution is deliberate and hard-won; it is not a "default" consequence of indecision, exhaustion, or cowardice. It is, as Bachelard described, the hallmark of "pure consciousness" and in this case, passive resistance of a quality understood by Ghandi and Martin Luther King, demanding extraordinary bravery and selflessness. Clearly, Delia is transformed and strengthened by the experience:

> Somehow, before sleep came, she found herself saying aloud: "Oh well, whatever goes over the Devil's back, is got to come under his belly. Sometime or ruther, Sykes, like everybody else, is gointer reap his sowing." After that she was able to build a spiritual earthworks against her husband. His shells could no longer reach her. AMEN. (138)

Later that night, when Sykes returns from his tryst with Bertha, kicking Delia across their bed and threatening to "mash [her in her] mouf," she goes "clear to the rail without answering him. A triumphant indifference to all that he was or did" (139). The "indifference" that Delia re-energizes through the Bachelardian rhythm of "work and repose" (29) bolsters her immunity to Sykes's control, while taking nothing from her. The "AMEN" that the narrator includes in this description marks the event as a highly personal, supra-religious growth experience for Delia, melding the "agape" complement to her nature with a Buddhist sort of transcendence—the paradoxical "half-virtue, detachment" (Bergson *Two Sources* 64) that reinforces her grasp on her own selfhood and her embrace of all that is precious and wholesome in the world around her.

In her second interior scene, trembling in the hayloft after escaping the diamond-back, the strength Delia gains through productive, recuperative duration is even more pronounced:

> There for an hour or more she lay sprawled upon the hay a gibbering wreck.
>
> Finally she grew quiet, and after that came coherent thought. With this stalked through her a cold, bloody rage. Hours of this. A period of introspection, a space of retrospection, then a mixture of both. Out of this an awful calm.
>
> "Well, Ah done de bes' Ah could. If things ain't right, Gawd knows tain't mah fault." (145)

The "calm" Delia achieves in the hayloft is the passage across the Jordan for which she has long been preparing. Like the climax of Leslie Marmon Silko's *Ceremony*, in which the protagonist restores balance to the universe by standing down the murderous taunts of his nemesis at a uranium-mine showdown on the autumn solstice, "balancing day with night, summer months with winter" (257), Delia's decision not to respond to Sykes's attempt to murder her in the "[d]og days" of August restores balance to her world, and she completes her passage to full personhood. The events transpiring during the period of approximately four months, spring through summer, crystallize in a pattern of seasons and cycles; this turn of the wheel marks a leveling season that carries Delia to a new passage in her life. She could not have traversed it without respecting and representing the forces of creation and destruction, equipped with the metaphorical harp and sword of Hurston's childhood dreams.

In consideration of the growth Delia achieves, it may be no coincidence that the flowerbed in which she lies after hearing her husband's death struggles is in bloom with the four o'clocks she had planted (146). While the flower's hour-name may also suggest some sense of maturation in the ripeness of a summer's day, the number four calls to mind Carl Jung's concept of the "divine quaternity" honored in pre-Christian religions (62), interpreted by Jung to represent parts in balance or "psychic wholeness" (Jaffe 273). Given Sykes's repeated vehemence toward "skinny" women and admiration for Bertha's "portly shape" (141), "skinny" Delia's achievement of full personhood may seem ironic. But for Bachelard, "the full and the empty, negation and affirmation, being and nothingness require each other" (Russell 5). Again, Delia is entirely capable of anger, but she knows how to use time to process and cast off the toxic excesses that can dull the sensibilities of a less mindful person. Sykes likes "portly" women because he can manipulate them, just as he did the snake, catching it when it had overfed on frogs (142). Sykes hates "skinny" women because he cannot entice, control, and enslave them. Knowing only the pleasure of short-term, corporeal satisfactions, he is fearfully mystified by a skinny woman whom fifteen years of abuse could not break, who stubbornly claims her right to act—or not act, by her own choice—rather than be acted upon. In so doing, she earns the privileges of a "definite person," as Bergson intended (*Time and Free Will* 165).

The song that Delia sings, with its refrain previously noted, serves not only as a mantra to invoke and extend her pure-time experiences, but also as a unifying motif: "Jurden water, black an' col / Chills de body, not de soul / An Ah wantah cross Jurden in uh calm time" (144). Water, representative of the feminine principle in Eastern philosophy, is clearly Delia's medium; she washes clothes for a living and diverts the intended effects of her husband's aggressions like a river against rock. Even more compelling is the notion of water as a cleansing agent. In this sense, the story's title highlights "purification" as a critical process step for Delia's pure-time generation; she "sweats" out Sykes's influences so that she can create an experience of pure time in which she might make decisions that are her own, untainted by his malice. That Delia emerges unsullied by Sykes's abuse is evidenced by her reaction to the sight of him crawling toward her, as the snake's venom is killing him: "[S]he saw his horribly swollen neck and his one open eye shining with hope. A surge of pity too strong to support bore her away from that eye that must, could not, fail to see the tubs" (147). The narrator's description resolves that Delia takes no pleasure or satisfaction in vengeance as she watches her husband die—neither her vengeance nor that of any perceived, higher power. She demonstrates only the nobler functions of human compassion and sympathy, for a fellow human from a shared origin, whose future possibilities are lost, irretrievably and unnecessarily. While it may be true that we cannot step into the same river twice, the Jordan River of Delia's gospel song represents the same river of time that offered the same prospects for improvement to both Delia and Sykes. The "cold river" that she imagines "creeping up and up to extinguish" his one open eye is the same "black an' col' river" that she crosses, by her own productive relationship with time, "in uh calm time." With each initiation of pure time, Delia makes choices that redouble her ability to invoke pure time. Similarly, with each treacherous act that he commits, Sykes redoubles his prospects of falling prey to his own treachery and dashing his own most favorable future prospects.

In this light, the disposition of both characters connotes the ancient image of Ouroboros, with the multiple meanings attached to it by so many cultures throughout the centuries. In Sykes's case, the story abounds with enough overtly phallic references to establish that the snake he brings home to accomplish what he could not—driving his wife out of the house—is an extension of himself. The Ouroboros model applicable to Sykes, then, is the compulsively insatiable serpent swallowing its own tail and thereby bringing about its own demise. Because Delia's actions come back to her with redoubled strength, her Ouroboros model seems more like the representation of the eternal return. More than a whimsical mascot for this essay, Ouroboros presents a useful framework for analyzing Delia's duration forays in an interdisciplinary context. Vonnegut, whose contributions to science far exceed those mentioned in the lines opening this essay, drew the image of Ouroboros in his 1973 novel *Breakfast of Champions* to answer the question, "What is time?" (200). If time is, as many conceive it, a curved continuum that folds back on itself to form a marvelous membrane, externally finite but internally fathomless, Ouroboros does as good a job as any more scientific blueprint of drawing our attention to the action inside time's borders that beckons our play. Far from a sealed orbit of repeated turns toward pre-determined outcomes, time inside the serpentine margin is a dynamic circuit, riddled with hidden chambers of possibility that expand in response to many variables, including human activity that is willful and creative, untainted by fear, malice, or greed. Bergson's model for duration, as further developed by Bachelard, is consistent with views of time held by some modern physicists. To Bachelard especially, as Russell notes, "mental energy [of the quality required to generate duration], given its complexity, is, of all forms of life energy, 'the closest to undulating and quantic energy', the most marked by vibration and rhythm" (8). In this respect, M-theorists, for example, may readily accommodate the action of duration generation as proposed by Bachelard and exemplified by Delia.

The all-important value to examining the philosophical notion of duration in the context of physical science, with a fictional subject like Delia as a theoretical case study, may be gleaned from consideration of "Sweat" as an allegory for the power that people may gather when they demand their right to make their *own* time and hence, their *own* decisions. Such a notion may be self-evident to people who experience time in ways other than the Western conception of a linear, future-directed impetus in which past, present, and future are irretrievably "agape." The fact is, though, that until more people are able at least to *imagine* themselves as free participants in time, capable of shaping futures different from those projected upon them by external entities' self-interests, our world is bound to remain perilously out of "round."

If the image of Ouroboros IS the shape of time, then its message to us may be the ultimate cosmic *koan*, challenging us *to look inside*, for therein lies the most marvelous action of time available to human experience, as fathomless as we dare imagine.

References Cited

Bachelard, Gaston. *The Dialectic of Duration*. 1936. Trans. Mary McAllister Jones. Manchester: Clinamen, 2000.

Bergson, Henri. *Time and Free Will: An Essay on the Immediate Data of Consciousness*. 1889. Trans. F.L. Pogson. Minneola, NY: Dover, 2001.

——. *The Two Sources of Morality and Religion*. 1932. Trans. R. Ashley Audre and Cloudesly Brereton. Garden City, NY: Doubleday, 1935.

Fields, Kerika. "Remembering Zora Neale Hurston During Women's History Month." *New York Amsterdam News* 4 Mar. 1999: 21. *Master File Premier*. Web. 18 Jan. 2001.

Fraser, J.T. *Of Time, Passion, and Knowledge: Reflections on the Strategy of Existence*, 2nd ed. Princeton UP, 1990.

Garcia, Irma. "Femalear Explorations: Temporality in Women's Writing." Trans. Eva Goliger Reisman. *Taking Our Time: Feminist Perspectives on Temporality*. Ed. Frieda Johles Forman. Oxford: Pergamon, 1989. 162–179.

"Henri Bergson." *Stanford Encyclopedia of Philosophy*. 12 Jul. 2011. Web. 12 Dec. 2011.

Hurd, Myles Raymond. "What Goes Around Comes Around: Characterization, Climax, and Closure in Hurston's 'Sweat.'" *The Langston Hughes Review.* 12.2 (1993): 7–15.
Hurston, Zora Neale. "Characteristics of Negro Expression." 1934. *Folklore, Memoirs, and Other Writings.* Ed. Cheryl A. Wall. New York: Library of America, 1995. 830–846.
——. *Dust Tracks on a Road: An Autobiography.* 1942. 2nd ed. University of Illinois Press, 1984.
——. *Folklore, Memoirs, and Other Writings.* Ed. Cheryl A. Wall. Library of America, 1995.
——. "Mules and Men." 1935. *Folklore, Memoirs, and Other Writings.* Ed. Cheryl A. Wall. New York: Library of America, 1995. 1–267.
——. "Sweat." 1926. *Literature: A Pocket Anthology.* Ed. R.S. Gwynn. 5th ed. New York: Penguin Academics, 2010. 136–147.
Jaffe, Aniela. "Symbolism in the Visual Arts." *Man and His Symbols.* Ed. Carl G. Jung. New York: Dell, 1968. 255–322.
Jung, Carl G. "Approaching the Unconscious." *Man and His Symbols.* Ed. Carl G. Jung. New York: Dell, 1968: 1–94.
Rackstraw, Loree. "Quantum Leaps in the Vonnegut Mindfield." *At Millennium's End: New Essays on the Work of Kurt Vonnegut.* Ed. Kevin Boon. State University of New York Press, 2001. 49–63.
Russell, Conrad. "Fictive Time—Bachelard on Memory, Duration and Consciousness." *KronoScope: Journal for the Study of Time.* 5.1 (2005): 3–19.
Samuels, Wilfred D. "The Light at Daybreak: Heterosexual Relationships in Hurston's Short Stories." *Critical Essays on Zora Neale Hurston.* Ed. Gloria L. Cronin. New York: Hall, 1998. 239–253.
Seedel, Katherine Lee. "The Artist in the Kitchen: The Economics of Creativity in Hurston's 'Sweat.'" *"Sweat": Zora Neale Hurston.* Women Writers in Texts and Contexts. Ed. Cheryl A. Wall. Rutgers UP, 1997. 169–182.
Silko, Leslie Marmon. *Ceremony.* New York: Penguin, 1977.
Vonnegut, Kurt. *Breakfast of Champions.* New York: Dell, 1973.
——. *Slaughterhouse-Five.* 1969. New York: Dell, 1991.

TALES OF TIME AND TERROR: WALTER BENJAMIN'S PHILOSOPHY OF HISTORY AND THE NARRATIVE AESTHETICS OF EDGAR ALLAN POE

Marcus Bullock

Abstract: Walter Benjamin's exploration of various ways in which he could incorporate literary and artistic modernism into a politically dynamic way of reading history led him to comment that Surrealism had fallen short of an original impulse in the avant-garde that he identifies with "the insights of Edgar Allan Poe." Yet the coherent thread that leads back to Poe's innovations does not lead forward to the idea that keeps returning in Benjamin's work—that disruption of the separation between the aesthetic realm and that of material relations would produce a future explosive emergence of truth. Poe constructed a form of literary language that sharpened the distinction between the fraught conditions of modernity and the intense discipline of literary composition. Benjamin sees the dialectical negation of that division between art object and commodity coming about by the destruction of the "aura" of a traditional work of art through its mass reproduction, by the supposed mode of apprehension by inattention among film audiences, and by the effect of Dada collages discrediting the traditional picture frame that "ruptures time." For Poe, it was clear that the preservation of that division remained essential to maintaining a standpoint of resistance to the social ideology that modern progress imposed on those caught up in its practices. We can perhaps now reconsider that capacity of resistant consciousness in confrontation with the further layer added to our contemporary array of distorted human relations, that of international terrorism.

Keywords: Avant-garde, Benjamin, crowds, Dada, historiography, modernity, narrative, Poe, surrealism, terrorism.

In Act IV, Scene 5 of Shakespeare's *Hamlet*, the king tells the queen, "When sorrows come, they come not *single spies*, but in battalions" (1905). The literary device of deriving a universal pronouncement from a particular state of affairs accomplishes more than deepening the sense of tragedy at that moment. It reaffirms the sense of literary time itself, the feeling of coherence that shapes isolated events into a body so that instead of a random series of shocks, we recognize an identity that links them together as a dramatic plot or a story. The "battalion" does not refer merely to the multiplicity of sorrows, but indicates that some principle has organized them by its implied connection. This conveys the essence of any plot or story, whether told in a dramatic or a narrative genre. Random shocks simply require that we cope or hurry on past them as best we may. An organizing principle, on the other hand, bears an implicit message of more to come. A literary telling of events communicates a task in the telling itself. That task takes the form of recognizing what will bring this passage through sorrows to an end. Fictional time does not permit its actors to go back and undo what has been done, but operates by a discipline of development that must look forward and reveal at the close of the story what it was they had not seen before. The story completes itself when the implicit necessity becomes explicit, as when, in a tragedy, the subject of a wrongful power or authority grasps why time will have purged it from the scene.

But the ability of a tragedy to provide the sense in which the flow of events is healed depends on the coherent link that runs between tragic events and permits them to speak to us as articulations of the story form. This provides the aesthetic assurance that draws us even to the most violent of tragedies. It invites us to set sail with a certain rapture upon a sea of troubles, because we can rely on a text. The structure of five acts promises us no reef will sink our ship until we reach the horizon of the story. But that is precisely where the realm of dangers in the arts does not imitate that of threats in life. When spies come among us to sow terror, they too wish it known they come not singly, but in battalions. They represent themselves in conspiracies that subvert all sense of calm expectation. This knowledge of an ill-will outside the horizon of our knowledge pains us far beyond the expectation that we will encounter sorrows, and many of them, within the normal passage of time. It is the abnormality of a hidden plan that torments us. The intention communicated in such acts takes possession of the future that it reserves to itself and inflicts on its victims. The cruelty of terror depends on its capacity to taint an absence with the image of an impending return. The secret of terror is to dominate waiting. To be a master of terror, therefore, is to be a master of time.

While changes in technology and organization have achieved grand endowments in material security compared to previous ages, the resources of tragedy as a means of reassurance in the face of suffering appear largely to have slipped from our grasp. Nothing proves more fragile than the confidence in time procured by the predictabilities of technology and economic rationalization. Capacities of artistic language, which for this purpose include expression in any medium, no longer enable us to create images of the world in which events, especially sorrowful events, speak to us as coherent voices. This leaves us vulnerable as perhaps never before to the distortions of anxiety. Karl Kraus, a literary lion in early twentieth-century Vienna, reflects on his reservations about progress and novelty in the striking expression "Ursprung ist das Ziel"—origin is the goal (67). The project of the future should not neglect the capacity by which we once endured a less determined existence with much greater resources of patience. Something akin to Kraus's feeling has begun to return in this century. That is to say, the ambitions of the future loosely grouped under the notion of globalization have also developed a contrary undertow. Having achieved so much in organization and technology, we who are caught up in those ambitions should feel that we are lords of all we survey, yet when we look for ourselves in times that lie ahead, we seem strangers there, if indeed we can picture ourselves there at all. Moreover, as we look back into our past, we see ourselves as strangers there too, so that, like Kraus, we need to take the past time as a project, a goal. The present in which we have stranded ourselves, a mere island of time, seems to many of us as though it might sink beneath our feet with the next movement of the clock.

By contrast, human memory sustains a quite different sense of its association with a generality of common experiences when fortified by the forms in which literary expression articulates the processes of patience. The possibilities of expression in fictions offer the ancient resource by which we understand how time permits us to knit up complex relations in events. This is exactly the enterprise that Kraus pursued throughout the enormous literary output of his long career, and which he elaborated in his monumental dramatic composition, *Die letzten Tage der Menschheit* (*The Last Days of Mankind*) and in the collection of commentaries *Der Untergang der Welt durch die schwarze Magie* (*The Destruction of the World by Black Magic*) in which the phrase black magic indicates the power of corrupted language.

In the discourse of terrorism, the avowed purpose nearly always invokes a return to some atavistic past when blood was "pure" and the nation "untainted" by modern complications. For Kraus, by contrast, the longing that he places in the past refers to the struggles against repulsive distortions of language visited on our understanding by nationalistic propaganda to sustain imperialistic warfare, by the crassness of journalism, and the power of anti-Semitism in Austrian politics. For Kraus, poetic language offers our truest and most open resource in the past from which we can return the present to a more generous and rational future. For this reason, the compressed formula "Ursprung ist das Ziel" stands as a direct contradiction to the atavistic origin dreamed by fantasies of purity and the diminished rhetorical space of "blood." The expansiveness of autonomous language needs no direct contact with action or with power to overcome the convulsive force of insecurity. Thus, the value implied in what this "Ursprung" means in the realm of fiction may not be apparent as long as the discussion focuses too closely on the instrumentalities of society and politics. Yet it does open up a significant alternative to the search for security through accelerated achievements in weaponry, technology of surveillance, and the social acquiescence that surrenders personal rights in order to further such material expansions. Kraus does not merely identify literary language as the repository of care in functional expression, and thus as a resource for a more thoroughly accomplished instrument—and thus, perhaps, a more effective diplomacy. The expression of relationships in fictions offers us a different model entirely. An aesthetic fiction in any medium or genre contains its origin in itself. That is what distinguishes a fiction from a fantasy and preserves the idea that it embodies its own modality of truth. And although it is habitual to speak of symbols in works of art, the language in which we might identify such meanings in an aesthetic construct should never be confused with the symbolic order of meanings in political culture. In Kraus's time, as in ours, the language of fiction had to struggle to preserve itself from the view that all discourse pursues a propagandistic purpose within ideological interests.

The sense of vulnerability produced by this reduction of articulation in discourse to the conflict of rhetoric and symbol obviously takes different forms today from those of the nationalisms of Kraus's time. He did not see what potential a phenomenon like the rather parochial forms of anti-Semitism in the Vienna of his day would produce in the Third Reich, nor the reach that would be achieved by today's sectarian hostilities. Its specific articulation of conflict enables international terrorism to reach higher levels of organization more easily and to persist longer. Its international component enables this kind of violent campaign to imagine itself as a war in an explicit sense, and it all too easily stirs its foes into treating the campaign against it as a war. Yet the difference enables these conflicts to last much longer than wars because they are not limited by the need to take and hold space, or to identify a goal that could be realized in any explicit time.

While the clear identification of a purpose, a policy goal, constitutes an essential part of what we traditionally understand as a war, the statement of the purpose itself does not classify a campaign of violence as a war. The purpose has to stand in a particular relationship to the means, and therefore in a particular relationship to time. The purpose of a war must have a role in the formation or preservation of an identity in the system of states and nations, and the means available in that purpose must possess a rational plausibility in its execution. If the purpose depends on a system of symbols expressed in the empty drama of violence, it can acknowledge nothing in an alternative system of value. Then neither side can discover a means of accommodation in the resources available to it, and the end comes only with the annihilation of one side or the other.

What defines terrorism as something distinct lies at least partly in the impossibility of the force of offensive violence producing a victory in itself. The terror in that case aims not at victory, but at the spectacle of blood and ruin as direct enactments of mythic identities and atavistic origins. A terrorist campaign can last for a generation or more because, unlike a war in the narrower sense, it does not have to show any kind of calculation about reaching a tangible goal in the form of a balance between cost and benefit. The cost *is* the benefit. As long as sacrifices are made—extracted in hurt to the target and to oneself—they serve the ritual purity of a ritualized origin.

The terror that citizens in the world experience when faced with the phenomenon of terrorism does not come primarily from the fear that it might touch them with actual violence. Anyone calm enough to count will know we are statistically more in danger from normal accidents or normal crime. The damage terrorism inflicts by spectacles of sacrifice produces a kind of nausea, a vertigo, because it operates as a negation, poisoning the plane of humane meanings. That is why such ideologies embrace these dark means to restore a dreamed past in a dreamed future. The conflict only confers the identity that violence seeks in terror when it draws a line of enmity without any speech that functions between the parties.

Nothing provokes quite as much dismay in the process of responding to international terrorism as the reverses endured by the political process and legal safeguards in the nation most securely endowed with the means to resist such self-denying fantasies. Among the nations of the world, the United States has exceptional clarity about the origin laid down in its founding documents. It stands on the original foundation of respect for the privacy and dignity of individual lives, among which the explicit abandonment of torture seems a most clear-cut component. Yet terror has so clouded this treasured past that we may be looking toward an America quite unlike itself in the future.

The sense of what stands in danger here grows especially apparent for someone who has joined American society as a conscious act and choice. If one is not an American by virtue of an origin in birth, one accomplishes becoming so in a process that may be unique in its completeness. The formal procedure of naturalization includes issuing a document to the new citizen stating that this identity has never been a matter reserved for blood or soil or birth or indeed nationality as understood in many other countries, since the United States explicitly declares itself not a nation-state. It was not established as a state to consolidate a territorial homeland for a pre-existing nationality, but was constituted solely as the idea that an as yet indeterminate future identity will grow from that origin. In this sense too its relationship to a past, and to an identity built on the foundation of the past, bears no resemblance to the national myth on which later postcolonial states built their independent status. The United States naturalization document affirms that one who shows commitment to join in the fulfillment of that idea of the future becomes as fully an American as anyone born a citizen, and that the collective identity of the United States becomes more itself for admitting all that a person of foreign origins brings from elsewhere in order to contribute to the idea that originates this union.

American history demonstrates a continual power of extension in the capacity of the state to articulate relations between communities through the integrity of that originating idea. What began in conflicts over slavery and the violations of aboriginal rights also created the means to overcome the deplorable elements of that origin. The founding documents have sustained every step of this tumultuous movement into its future. The integrity of that inheritance permits us to find ourselves coherently in time.

The individual process of becoming an American citizen thereby unfolds as a representation of how the union became a country with its own distinctive tradition formed within the inheritance of that idea and the language of our originating documents. That is exactly the point made to new immigrants in the document that describes the status they have now entered. Nothing that they bring with them from a previous heritage contains the symbolic force to exclude them; on the contrary, the differences they bring to it "make our country not less, but more American."[1]

---

[1] This is quoted from the message of welcome handed to all new citizens on the occasion of their swearing in. It takes the form of a letter, sixteen lines in length, without any departmental designation, but identified merely by the signature of the President, Republican and Democratic alike, describing the immigrant's new status. It takes one's breath away at such a moment in one's personal history because it is so welcoming of individual specificity. Its definition of citizenship connects the process of naturalization to the original relationship between state and people and invokes a future whose openness similarly unfolds the idea on which the country builds its evolution. This process cannot

That is to say, the principle of law on which the Constitution stands establishes those spaces of discourse into which citizens may enter and represent themselves rather than submit to their being represented. Yet nothing guarantees that this will always continue. The facts of American history also are replete with shocks and outbreaks of insecurity in which the country has not grown more, but less itself.

The question of the character of the language in which one is to remember and communicate the origin of that idea, however one came to participate in it and speak for it and within it, whether by birth or immigration, can now appear in this form: how does the country continue to tell itself the story of that idea in its integrity so that it does not lose what it has been? History may consist of facts, but these do not speak for themselves. The practice of historiography has an inventive component. For the Greeks, the writing of history had its own muse, Clio. That implies the recognition that the writing of history drew on the powers of imagination shared with its sister arts for its capacity to weave time into a guiding thread through the labyrinths of change. Yet the ninth section of Aristotle's *Poetics* offers a more restrictive view. He thought poetic expression "more philosophical" than history because it dealt with "universals," whereas history dealt with "particulars" (12). It has become harder and harder as we become more and more aware of differences between literary traditions and the cultural conceptions on which they draw to see philosophical truths in poetic expression. Repeating formulae that invoke the systems of universals from so long ago now must rank as more of an idle consolation than a serious confrontation with our situation today, stranded between darkness ahead and darkness behind. The modernity that Baudelaire recognized as so epoch-making in the figure of Edgar Allan Poe has ushered in a quite different function for aesthetic objects than such a "philosophical" content. The universal truths once expressed in a heritage of myths in classical genres now survive in frames of distance and irony. The representation of disruptions and fragmentation predominates as the intimation of how we now experience our mortality. Aristotle's formulation indicates that he sees a social value in fictions because they provide us with representations that transfigure the meaning of a single moment in time into a permanent illustration of ideas from which we learn timeless truths.

be reduced to an "ideology" in the commonplace sense in which an enclosed vision denies the reality or the truth in other ways of seeing. It is quite the opposite. It is the idea of development itself.

That is part of the pleasure that Shakespeare affords. His characters formulate the patterns in which life seems restored to familiarity even in the midst of suffering. Yet every stage of modern development—from Baudelaire and Mallarmé's acknowledgment of Poe, through Rimbaud and the avant-gardism of the twentieth century, to Pound, Joyce, Beckett and the repeated phases of radical artistic autonomy—recognizes that aesthetic authority no longer accrues in the form of an edifying moral authority to such lessons in general truth. The resources of fiction now serve to order the disturbances of fact according to a different principle. The deep universal observations by which Shakespeare's poetic speech entered our everyday English evade our sense of truth when modern life is fraught with unrhyming events. Some writers still offer edification through representations of the typical, but those with powerful invention explore our experience of the exceptional. Whatever the lingering charm of literature formulating generalities of worldly observation and moral worth, a different urgency has arisen now by which we judge the seriousness of literary art. There is something else in our modern experience that artistic values need to confront in a "philosophical" manner, namely the shock of deep unfamiliarity. And this is why Poe insisted so vehemently in his essay "The Poetic Principle" that the essence of true poetic creation lay entirely in constructing its own disciplined composition, and preserved its artistic authenticity by that autonomy. No value accrued to it by representing social values of a worldly character, which was very much part of how many of his contemporaries had engaged the moralistic philistinism of a reading public. Therefore, he made no bones about declaring that literary art had no business involving itself with "the heresy of *The Didactic*" (700) and had "no concern" with "Duty" or "Truth" (703).

What remains as common ground between aesthetics in Aristotle's world and our own is the shared project of ordering time. As in Aristotle's era, the work of art in any medium or genre identifies itself as a closed entity, a text, a performance, an object framed and set apart. All such expression therefore maintains that same principle of integrity. That which belongs to the work, and which the process of composition brings into being, is set off against that which does not. It has, just as Aristotle put it with such deceptive simplicity, a beginning, a middle, and an end. This relation to time runs unchanged through the history of the arts. Music, as before, casts off into a flow of its own time and comes to a close of its own defining, and as before, painting and sculpture lift an image out of the stream of disordered events, the series of successive present moments fading before we can grasp them right, to preserve that "still" image for us no matter how many centuries have passed outside the frame. Lyric poetry captures the mood of time pausing; a story organizes time passing; and the epic represents the fullness of worldly time returning us to the origin of things. Yet in every such genre, aesthetic creation has grown different because the experience of the present and the demands of time that we must now accommodate have also grown different.

For this reason, it should not strike us as strange or strained to look anew at the practice of historiography as *musisch*, in that telling German term—as joined through its muse with the spirit of the other arts. History as the sequence of chaotic unique events might correspond to that realm which is less philosophical in Aristotle's opinion than poetry, but historiography is a form in which the process of writing knits up the past according to general conceptions of human motivation and human relations. It has become a commonplace observation since Walter Benjamin explored the idea in his essay "The Storyteller" that the large patterns of narrative technique in the history of history writing change in parallel with evolving narrative techniques in the literary history of fictional stories. He found it highly significant that the great works of nineteenth-century historicism structured the past in ways that mirror the grandly organized nineteenth-century novel's presentation of its plot. Facts and fictions may constitute opposite domains in the obvious sense, but in the wider realm of the means at our disposal to construct coherent relations in the representation of time, we solve equivalent problems of composition when we order time both in historiography and in the telling of fictions. Therefore, both may function as complementary means by which we may secure ourselves against the disorienting impact of terror, of violence that aims to call the foundations of our identity into question by diminishing the place of language.

The fourteenth of Benjamin's eighteen theses in "On the Concept of History" ("Über den Begriff der Geschichte") quotes Kraus's line "Ursprung ist das Ziel" as its epigraph (395). While this conception of a future taking us back in time might seem a natural growth in the mind of a conservative thinker like Kraus, Benjamin's text blazes with the passions of revolution. Yet neither with regard to the future nor the past do differences that separate conventional ideology divide men joined in their understanding of the dangers posed by diminished capacities of language. Benjamin found no difficulty in appreciating the powerful moral and political criticism Kraus directed against the terms of progress. The concept of a revolutionary future that Benjamin develops in this philosophy of history also meant retrieving and restoring a grasp of human relations in a language that dissolved or disrupted the dark spell of a mechanistic chronology.

The relationship between these two thinkers naturally goes beyond this quotation. In 1931 Benjamin wrote an extended essay articulating their common suspicion of progress. It would have startled neither of them to watch naïve hopes placed in technological advance and liberalization stumbling so painfully today. In 1940, when he composed the pungent theses we know as "On the Concept of History," Benjamin's previous positions on class domination and a future in revolution had undergone an enormous shock with the Hitler-Stalin Pact of 1939. As a stateless refugee in France, his situation was already fraught with urgency. With the first defeats of the war, his remaining time seemed threatened with dangers it would be ever harder to escape; with his suicide under arrest on the Spanish border, that time soon ran out. Those theses accordingly show a dazzling compression of thought. Never before had Benjamin been compelled to reach so radically into the philosophical past of theological and Messianic thinking. The arcane images and sudden illuminations by which he presents models for the activity of "the materialist historian" resemble nothing else in the philosophy of history.

The principal idea that he foregrounds as a solution to the new level of problems goes back to his 1929 essay "Surrealism," where he closes with the idea that the moment of intense alarm ushers in the opportunity to master the present. The mastery to be achieved in "On the Concept of History" involves bringing time to a standstill and subjecting it thereby to a "shock" that causes it to crystallize and open itself up to analysis. What he means by shock here is not immediately open to analysis itself. For that reason, it has invited casual commentators on his work to apply the term in any way to which its dramatic sound suited it. The term "shock" in his essay on film technology, "The Work of Art in the Age of Its Technological Reproducibility," expands the difficulty even further because here film delivers a shock to the viewer; nothing permits viewers to "crystallize" the film by bringing it to a standstill. Benjamin's uncertainties about the relationship of aesthetic time to material time, however, go back to the way he attempts to contrast what he calls "the insights of Poe" with inadequacies in Surrealism.

Reception of "The Work of Art" essay presented difficulties from the beginning, some of which Benjamin attempted to deal with in a second version. This introduced some important changes, but in both forms the experience in making meaning he ascribes to the film public differs from what he explores in the experience of making meaning by a materialist historian in "On the Concept of History."[2] A study of past records by an individual mind motivated by intense urgency clearly corresponds to the work he himself knows, whereas the behavior of a mass audience that purchases entertainment before a public screen does not. In both versions of "The Work of Art in the Age of Its Technological Reproducibility," he praises the possibilities of reception by an audience in a state of distraction or inattention.[3] He compares it to the means by which public use of a building mediates familiarity with it because "[e]ven the distracted person can form habits" (120). The debates over the significance of "shock" set in motion by the 1969 edition might have developed less intensity had it presented Benjamin's modified text. In any case, both versions more consistently apply terms like "casual noticing" (120) and "expert appraisal" (116) to describe film reception rather than "shock." In the first version, coping with shock appears with a recommendation for mental activity closer akin to the dynamism of a materialist historian. There he observes that "the shock effect of the film ... like all shocks ... seeks to be buffered by heightened presence of mind" (*Illuminations* 238). In the later version, that line has been deleted from the main body of the text, though it may still be found buried among twelve pages of densely printed endnotes ("Work of Art" 132).

In his essay on Surrealism, on the other hand, we see Benjamin uniquely equipped to offer powerful and disciplined insights into this phenomenon. His discussion adopts a manner of reading the literature and commenting on the artwork of Surrealism that needs no recourse to the Messianic claims about historical truth or political redemption in "On the Concept of History." Similarly it has no need for the speculations about an experience he did not share when, in both versions of the "Work of Art" essay, he argues that the film-going masses imbibe a progressive critical consciousness through the distraction this medium affords. He interprets the Surrealist movement, still a vigorously contemporary phenomenon at the time,

---

[2] Christoph Zeller has recently published a very helpful essay with much to say on this issue. He offers insight into the reactions of Benjamin's contemporaries to the essay, follows the subsequent discussion up to our own times, and provides a sophisticated assessment of the position the project takes in the larger picture of Benjamin's ideas in the writings that preceded and followed it.

[3] The 1936 version, chosen for the larger and more authoritative Harvard *Selected Writings* in 2002, contains an important change from the 1935 version that appeared in the earlier collection, *Illuminations*, edited by Hannah Arendt, published in 1969 (translating the essay's title with the somewhat more relaxed English "The Work of Art in the Age of Mechanical Reproduction").

within a longer tradition of literary modernity. Rather than establishing its political relevance to the struggles of the day directly through the Surrealists' embrace of Communism, he does so by reading the movement through its origin in the hyper-aestheticism of the nineteenth century. The inspiration of a hermetic aesthetics for Benjamin the historical materialist lies in its ability to emancipate itself from the conformities of a pragmatic world fascinated by the lure of progress and thus prey to domination by the powers that control and propagate it. Tracing the movement back into the previous century through Rimbaud and Baudelaire, he remarks that Surrealism might "sink its roots deeper" than its current dogmas "to the depths of the insights of Poe" ("Surrealism" 215). The insight that distinguishes Poe beyond all his contemporaries and has established him as a permanent influence among Europeans beyond all other Americans is his disciplined insistence that literary composition should demonstrate an absolute commitment to its own inner coherence and nothing else. Only this discipline in expression can equal the shocks of savage irrationality that modernity imposes.

Though most literary critics with his political concerns had viewed such esoteric aestheticism on the part of the Surrealists as reactionary or decadent, Benjamin contradicts this opinion as the "obligatory misunderstanding of *l'art pour l'art*" ("Surrealism" 211). He interprets the Surrealists' autonomous practice of art as a radical assertion of freedom. The insights of Poe emancipate modernity from obligations toward the visible achievements of progress and the institutions of liberal politics that had taken up so much of his contemporaries' literary energies. That is Poe's point in "The Poetic Principle" when he objects to "the heresy of *The Didactic*" (700) and asserts that poetry has "no concern whatever with Duty or with Truth" (703). Benjamin adds a justification for his own position on *l'art pour l'art* in the Surrealism essay with the comment that "art for art's sake was scarcely ever to be taken literally; it was almost always a flag under which sailed a cargo that could not be declared because it still lacked a name" (211–212). Revolutionary historiography, as Benjamin conceives it in the fourteenth thesis of "On the Concept of History," forms its understanding of the present by making what he describes as a "tiger's leap" into the past. It has flair like that of fashion seizing on the "topical, no matter where it stirs in the thickets of long ago" (395). The medium of historical development that links origins and futures in Benjamin's view clearly supersedes any simple succession. Indeed, the fifteenth thesis most emphatically values the way in which societies and cultures recollect the dates of defining past events through the year in calendrical time but rejects the way daily activities are regulated within the empty linearity of chronological time.

The negative view of time measured by clocks, along with the action of similarly oppressive clock-like devices, locates a point of correspondence between Benjamin in these theses and Poe in such stories as "The Masque of the Red Death," "The Scimitar of Time," and "The Pit and the Pendulum." In the fifteenth thesis, however, Benjamin goes further than this representation of time in terms of aesthetic sensation; he finds a connection between the horror of a mechanical movement in time and a historical phenomenon that locates it in the political sphere:

> Thus, calendars do not measure time the way clocks do; they are monuments of a historical consciousness of which not the slightest trace has been apparent in Europe ... for the past hundred years. In the July Revolution an incident occurred in which this consciousness came into its own. On the first evening of fighting ... the dials on clock towers were being fired at simultaneously and independently from several locations in Paris. (395)

It is precisely for its complicity in this failure of consciousness over the previous hundred years that Benjamin's theses collectively criticize the scholarly practices of "historicism"—the term by which he designates the grand manner that had come to dominate historiography in the nineteenth century and owed so much to the fictional worlds of the nineteenth-century novel. Both narrative practices rely too heavily on recounting time as a chain of events ruled by a rigid set of stages. This produces both the contrived "meaning of a life" in the conclusion of the novel ("Storyteller" 155) and justifies the present as an inevitable result of a logical process in the writing of history. The order of time within which historicism projects the inevitable movement of progress, mistakenly and for reasons of political conformism, resembles the chronological time within which the natural sciences pursue the unvarying laws of cause and effect. "Historicism contents itself with establishing a causal nexus among various moments in history," Benjamin notes, "[b]ut no state of affairs having causal significance is for that very reason historical" ("Concept of History" 397). That is to say, the impulse to mount the enterprise of the future solely in terms of a rationalistically conceived progress completely denies us the means to understand ourselves historically. And this in turn impedes our capacity to identify our real interests as members of a society making decisions in the political sphere.

The alternative he proposes establishes a freedom of movement that opens the whole of time to the establishment of connections, making it the goal of historiography to recreate history as, so to speak, our homeland. In "The Storyteller," he opposes this conception of history reinstated as our homeland to Georg Lukács's notion of the modern condition as "transcendental homelessness" (41). Were the language of political and historical representation recreated in the way that Benjamin imagines it might be, neither the attenuations and obscurities of the economic system, nor the atavistic eruptions of violence that finds its symbols an attractive object for attack, would render us incapable of resisting their oppressions of our historical consciousness. However, that recreation, as a Messianic act, remains only a thesis, albeit one rendered with a certain magical force in Benjamin's alluring prose.

The critical force that Benjamin manifests in dynamic juxtapositions modeled on Surrealism or on "the insights of Poe" acts with more philosophical than political capacity. He offers a vivid rejection of history conceived as a system of natural processes because these by their nature—ruled by causality and bound by natural laws—lack a goal. They cannot express the pursuit of a human value or human identity. Where we value an origin as our own, as we do when we identify with our citizenship in a national community or our participation in a civilization, that origin defines the time that flows from it as a unique medium of connection to our future. Our origin establishes an identity that comes before a beginning, and a goal implies a condition that falls to us as the possession and fulfillment of a desire after the process has reached its end. The distinction between these conditions constitutes a radical separation within stages of history and possibilities of knowledge about it. Human action arises in the context of a language that precedes it, and the results of that action only permit articulation by a point of view that looks back on it. Thus the concept of revolution for him preserves a meaning still visible in its etymology—a rolling back of circumstances to remove an intrusive growth of injustice.

The historian most dynamically engaged in the present is also the one most certain that the active moment of political response is what constitutes as well as reads history. Thus the fourteenth thesis of "On the Concept of History," immediately following the epigraph on origins, opens with the distinction by which history raises the notion of time to a higher value: "History is the subject of a construction whose site is not homogeneous, empty time, but time filled by now-time [*Jetztzeit*]" (395). In the eighteenth thesis, where he excludes the realm of cause and effect from history, Benjamin explains that only the engagement of this dynamism in the present accomplishes the tiger's leap into the past. Only by virtue of the intervention of the present will a fact in the past become "historical posthumously, as it were, through events that may be separated from it by thousands of years" (397).

What Benjamin insists on as the truth (indeed, the Messianic truth) of history, assembled in such leaps and bounds, depends on the active focus of the time so constituted. The ability to seize on a judgment that interprets a historical correspondence happens in the instant of "now-time" with the suddenness of an artistic inspiration. It identifies what historiography has in common with the arts, and why it has a muse of its own. While we would incline to say that a discipline of knowledge dealing in causal relations *finds* the truths it propagates, and the arts make their truth, we see a dangerous contradiction afflicting Messianic history. If a causal relation can be shown to be false, then the process did not find what it claimed. If a work of art is poorly made, then it does not make a coherent form for which it achieves aesthetic assent. The political judgment of history is true for those who adopt it as their form of engagement, but merely an assault for those who are to be removed by that tigerish decision. The example Benjamin offers us for our assent in that thesis is, of all people, Robespierre and his politics of terror. The form and the completeness in which we can apprehend truth through a fictional object depends on the fact that it is not a fact. Nothing of it spills over beyond our purview into realms hidden by the worldly horizon; because it is explicitly a made object and entirely so, the privileged frame of the work reveals its content in entirety and, as it were, on all sides. It does not invite a practice of terror on one side against the other.

Within the dialectics by which Benjamin distinguishes between aesthetic and worldly experience, the frame of a work of art does not merely mark off a spatial limit but also a temporal one. A great deal of his dramatic objection to "cultural treasures" in "On the Concept of History" has this determination at its root. The work of art not only generates a domain exterior to that flow of time in which historical decision would open its potential to us, but it thereby also fosters the continuous hold of usurping powers and injustice. For this reason he claims that "[t]here is no document of civilization that is not at the same time a document of barbarism" and that for critical historians, "they have an origin which he cannot contemplate without horror" (392). The formal closure that distinguishes such objects of contemplation certainly transfigures the time that the listener to music, the reader of a book, or the observer of a painting takes for these pursuits as opposed to those of ordinary life. And despite his intense preoccupation with aesthetic autonomy in the heritage from Poe, via Baudelaire and Rimbaud, to the twentieth-century avant-garde, Benjamin still objects to what he regards as a loss suffered in our apprehension of worldly interests as a result of the attraction and fascination with which we contemplate autonomous aesthetic works.

Accordingly, he looks for phenomena in the development of avant-gardism itself that break the spell of all aesthetic fascination. In his 1934 essay, "The Author as Producer," he cites Dada collages of everyday objects as important for their resistance to the conventional reception of a framed picture. They invite the viewer to approach them with expectations derived from an older heritage, but without finding the "aura" of a traditionally privileged creation. This exposes the function of the frame by which the Dada object identifies its relation to traditional artworks. Because the production of these objects no longer repeats any of the expressive functions that communicate a traditional artistic content, they operate by reframing the conventional reverence toward a work suffused with what he criticizes explicitly as "eternal value" in "The Work of Art in the Age of Its Technological Reproducibility" (101).

In that essay, Benjamin welcomes the idea that photographic media will obliterate the "aura" that we experience in the presence of individual, original great works of art because the aesthetic power vested in such objects interrupts the consciousness of concrete, material, lived time and substitutes a distanced, atemporal apprehension instead. "The Author as Producer" claims that when Dada artists framed objects gathered essentially unchanged from worldly detritus, "the public was shown: Look, your picture frame ruptures time; the tiniest authentic fragment of daily life says more than painting. Just as the bloody fingerprint of a murderer on the page of a book says more than the text" (774). In this way Benjamin finds a name for that secret cargo carried by *l'art pour l'art*. That name is artistic counter-aesthetics.

The dialectical practice of history takes a cue from the shock effects achieved by avant-garde artists in proposing a kind of counter-history. The historical materialist must "blast open" ("Concept of History" 396) the continuum of historicist time and thereby bring the deceptive movement of progress to a standstill. He calls the resulting isolated instant a "Messianic cessation of happening" in the seventeenth thesis of "On the Concept of History" (396). Yet neither in the arts nor in historiography have such contraries provoked the divine to intervene in politics so far. Nonetheless, that contradiction itself might prove a more vital source of the very insight that Benjamin would like the Surrealists to trace back to Poe. The method of Poe's composition lives on as an inspiration to the work of the present, not as an item of prosecutorial evidence and authority to which the present submits.

The effect achieved by Kurt Schwitters and other practitioners of art that includes objects of daily life within the frame did not disqualify them from ultimate inclusion within a high tradition of the artistic heritage. Their productions too become objects of rapt contemplation, displayed with undiminished privilege in private collections and in public museums. Marcel Duchamp's famous urinal has been treated as a great original work (though the actual "original" first signed in New York was lost and its replacement destroyed in an accident). Now it is cited with reverence as a classic of modernism. Benjamin's invocation of a Messianic moment as a complete and restored dimension of worldly truth renders the concept of truth itself obscure. Here we should apply Benjamin's advice to "brush history against the grain" ("Concept of History" 392) to his own interpretation of these phenomena. His dialectical history of aesthetics producing its opposite in Dada reveals, contrary to his assertion, the undiminished power of the frame. An original collage by Schwitters radiates the spirit of its moment. The object holds the attention of the viewer who stands before it, and cracks a fault-line in worldly time within which the moment brims with the alternative aesthetic time. "The Work of Art in the Age of Its Technological Reproducibility" claims that "[b]efore a painting of Arp's or a poem by August Stramm it is impossible to take time for concentration and evaluation, as one can before a canvas of Derain's or a poem by Rilke" (119). Yet history demonstrates little of that difference. The paintings inspire the same necessity to visit, to return, and to experience the transforming sense of their presence, and this same relationship to time holds for the literary work too.

Informative or explanatory texts quickly lose our interest with mere repetition, whereas a work of art holds us in a uniquely rapt relationship in the first place, and draws us back to it with a unique attraction ever afterwards. This returns us to the question raised at the beginning of this discussion. The aesthetic time which we can repeat, and in which we can linger, offers the only reliable antidote to the sense that our time is not our own, that we are strangers here, and that we do not find ourselves any longer in an origin to which we look back, or a future into which we peer even two steps before us. And the sense in which we can find ourselves again in such a possibility of return formulates what we mean by a truth of fiction. The fictive events in a literary narrative are not "untrue" in the sense that they are deceptions or errors, but only in that they have been assembled in a constructed time outside the realm of "facts." Fiction unwinds the route from the end to the beginning that already contains it. Poe declared in his essay "The Philosophy of Composition" that "every plot . . . must be elaborated to its dénouement before any thing be attempted with the pen" (675). The capacity of such coherence to sustain us in the face of deep terror, in times of radical insecurity like our own, marks the true value of that esoteric aestheticism by which Surrealism could "sink its roots . . . to the depth of the insights of Poe" ("Surrealism" 215).

The alternate historical standpoint that Benjamin proposes in the theoretical power of a Messianic cessation of time is supposed to raise ordinary objects to such an intensity of presence that they somehow speak for themselves. This theory also has its counterpart in the method of fragmenting a text in order to apply interpretations to isolated material cited thereby, and of turning these to run counter to the apparent tendency of the text as a whole—a method also conveyed in the phrase "lecture symptomale" associated with Louis Althusser.[4] But isolated fragments do not speak for themselves; they speak for the person who has detached them from their original place. Such a process of critical commentary corresponds to the effects achieved in film editing by the process of montage to which Benjamin attaches such primary importance in "The Work of Art in the Age of Its Technological Reproducibility." The fragments are joined by productive ideas equivalent to tigers' leaps. Here we see Benjamin take an enormous additional step beyond the phenomena he observes. The "Messianic cessation of happening" is supposed to be the opportune moment at which Benjamin's historian will "blast a specific era out of the homogeneous course of history—blasting a specific life out of the era or a specific work out of the lifework" ("Work of Art" 396). But the Messiah can be relied on neither for his arrival, nor for his beneficence. Where he is silent, the noise of Benjamin's blasting would simply leave us with

[4] Althusser develops this radically inventive but questionable procedure in his book, *Pour Marx* (*For Marx*), written in 1965. There he attempts by this means to derive an unassailable philosophical knowledge from the writings of Karl Marx. For the purposes of this essay, the further development of this method in work analyzing and critiquing Althusser is more useful. See, for example Samuel Solomon's recent essay (5). Antonio Negri describes how close to Benjamin Althusser comes in this manner of reading. He concludes, "Combien cet Althusser est proche de Benjamin!" in the opening section of his essay "Pour Althusser: notes sur l'évolution de la pensée du dernier Althusser," which he makes available on the internet at http://multitudes.samizdat.net/Pour Althusser: notes sur l'évolution de la pensée du dernier Althusser.

bombastic political declarations in historiography and a residue of didacticism in literary interpretation, were it not for all the subtlety of his prior exploration of the fault-lines between the aesthetic, the political, and the didactic functions of any text or object.

The point at which we should mediate between the apparent esotericism of Benjamin's thinking in these declarations, and the concrete demands made of us in the worldly human condition, does not lie anywhere near his vague invocations of class struggle. The point from which he accomplishes that sudden intensification of vision has its origin most typically in experiences conditioned by the modern commercial metropolis with its unceasing flow of the urban crowd. He recognizes this significance of the crowd in Baudelaire's poetic journey through Parisian streets under the impact of the industrial era, and through Baudelaire he recognizes the immense importance of Poe for all literary modernity. In the essay "On Some Motifs in Baudelaire," he calls Poe's story "The Man of the Crowd" nothing less than "the classical example among the older versions of the motif of the crowd" (324) and likens the narrative technique by which Poe framed an essence of this new experience to the new invention of the camera. This device "gave the moment a posthumous shock, as it were" (328), when it held an instant of time motionless and thus available to subsequent critical absorption.

Poe's tale relates his narrator's continuous trailing over many hours of an old man whose relentless life energy seems to derive entirely from his presence among crowds. The narrator only gives up this investigation when the man's wanderings bring him back to his starting point. The story concludes with the comment that we should not make any further attempt to penetrate this unnatural ceaselessness of motion other than to recognize that "this man is the type and the genius of deep crime" ("The Man of the Crowd" 239). We see no indication that he has broken any laws, only that his existence requires no connection to the realm of law. The story is deeply fictional in the boldness with which it portrays a figure who could not exist, and it describes a host of fleeting figures in the crowd according to characteristics that are only a little less dubious. Yet it also captures a vision of London crowds that has fascinated readers since it first appeared. This is possible because, in the slightly feverish sensibility of the narrator and his entranced and appalled response to the Man of the Crowd as an alien expression of energy, the tale portrays an individual sensibility in the absolute state of rapt discovery. He looks into what should be the familiar motion of the crowd flowing through public spaces, and loses his gaze in the bottomless darkness animated by some kind of vital current that he can never see except in its incomprehensible effects.

In this discussion of Poe and the crowd, Benjamin also cites an observation by Heinrich Heine regarding the "horror" that one feels when confronted by the constant movement of a great urban concentration of energies; and, by way of contrast with Poe's modernity, he cites an earlier story focused on urban crowds as recounted by the Romantic writer E.T.A. Hoffmann. What distinguishes Poe's drastic modernity is the vanished sense of the eternal as a support and background of time. For Hoffman, the idea of market day, an expression of calendrical time, preserves the framework of rhythms evoking the canonical week and the realm of nature. These are dimensions that fill repetitions with a ritual quality, and with an idea of patterns of life that also inscribe the law within them. By contrast, Poe's crowd and the man drawing energy from it circulate unceasingly in an empty, homogeneous time. The constant panic on the part of this strange figure, whose life energy flags whenever separated from the flow of passers-by, reflects the unrelieved precariousness of that existence, condemned to find his way onward through a time without a quality other than this motion.

The horror that gripped Heine, and the fascinated alarm that held Poe's narrator, convey to us the sense of a newly encountered flowing void. The power of that impression lies in its capacity to enfeeble a sense of our own original time. The movement of the crowd, having no identifiable origin or goal, leaves us bereft of time, motionless and passive amid a rush that belongs to everyone else but not to us. Our own modernity has in no sense solved or overcome the quality of such experience. On the contrary, our depleted experience leaves us with a bewildered emptiness of expectation that the practices of terror can all too readily captivate.

The BBC program "The World at War" noted the eerie significance of a stratagem of psychological warfare at the Battle of Stalingrad. After a huge German force had been cut off and surrounded by the Red Army, Stalin had his generals set up loudspeakers at the periphery of the German-held ground, and in addition to the usual messages of challenge, threat, or mockery, they played the sound of a ticking clock.[5] This sound did

[5] The story of techniques of intimidation on the battlefield must be as old as warfare, and the conflict on the Eastern Front in World War II seems to have been especially rich in inventions on both sides. The Soviet Union used sentimental songs about Germany and home to dampen martial spirits, announcements of the numbers in which Germans were dying, even tango music, as well as the sounds of the ticking clock. See Beever (307–308); further see Craig (286).

not depict what lay ahead for the enemy, or when it was coming. It just marked off the relentless emptiness of the time remaining to them before the assault would be renewed with even greater force.

The image from Stalingrad reminds us that in the horrors of war, terror competes with terror. Hitler and Stalin compete in horror ranged against horror. Each struggles to impose the same power over the other, and to enfeeble an opponent under the spell of terror. Yet quite clearly we need another form of mastery, a constructive process by which to rebuild the fullness and completeness of time that terror taints, oppresses, and reduces. The essence of that which makes us feel overmastered lies in a sense of blindness veiling our original future. An opponent plans and chooses the moment for which we must wait. We can escape from that state of waiting by delivering a shock to our foe. Answering terror with terror, we commit ourselves to delivering a blow that our opponents do not foresee. We will have taken something back, we may think. It may feel in our frenzy of anger that we have retaken the landscape of time for ourselves, but the origin of that anger was not ours, but theirs. We have only given way before the pressure of a time ruled by their act of violence to which we merely react. Thus we will not have regained the mastery of time that our enemies took from us. If we succeed in this struggle against our foes, we master their means only.

Now all of this lies well within the realm of common knowledge. Citizens of the United States and all other countries that confront organizations endangering their security deal with constant reminders of a danger. Whatever the real likelihood of being affected directly by an actual event, everyone is aware that the United States Department of Homeland Security maintains a code indicating threat levels with colors ranging from red to green, and if the news media do not report it, one can always go to the Department's website to see the color of the day. This creates our common longing to return to the green zone. Green signifies a normal state of our being. Consoling as it is to put a label on anything, it is not very thoughtful. And wherever we long for a "normal state of our being" without identifying clearly what that means, we have abandoned thought for a longing mastered by dread. We might then ask ourselves: what is there about the ticking of a clock that is so profoundly unsettling, even while the mechanism also signifies normal time? It is certainly not "normal" to be waiting for an attack. Yet we are always vulnerable to something that time can bring. As the anecdote from the July Revolution and the gunfire aimed at Parisian clocks suggested to Benjamin, the oppressiveness of mechanized time inflicts an inhuman relentlessness on workers who experience the subordination of quantified labor. For them, the clock is always ticking. Time is always running out. But time also offers itself as a vehicle by which natural disasters reach us. In that realm, time will always bring catastrophic events, the tsunamis and earthquakes that find even the best organized nations unprepared. Nonetheless, such random events belong to the normal course of nature. Even though they far outstrip the worst attacks of terrorism in their effects, their possibility intrudes on our minds quite differently. Terror arises with the cruelest acuity where we know that our time has become someone else's domain.

And if the minutes sometimes fly by, and sometimes slow to a crawl, or even, in moments of deep serenity, seem to rest, the clock is always ticking somewhere. Anticipation of natural events regards them as disconnected, as "single spies," and we counter them with the human spirit of hope. The awareness of cruelty, on the other hand, requires of us a distinctive form of discipline to resist. Benjamin makes a political point of what happens when we think of history as a natural process. We can choose to make a distinction between a flow of time that we call normal and one fraught with alarms, and we may succeed in charging that distinction with the greatest importance. It depends on what we see and what we do not see in our future when we glance backward at the past. We feel we are in normal time when we interpret what we see in the backward glance as a system of repetitions that will proceed in a continuous flow. What we then call the future simply shows a mirror image reflected in the calm pool of the past. But we need to ask ourselves: an image of what? What is it that has the power to attract and reassure us so that we come to regard it as "normal"? Benjamin points out in "On the Concept of History" that this does not constitute a true future at all. Imagining a serene history that merely unfolds its causes and effects according to a transparent logic laid down in its apparent origins does not, in his view, constitute a past either. This view of time not only rationalizes and justifies the unjust conditions of the present, but also endeavors to retrieve and dwell in the past as a substitute for the present in which passive historicists can make no moment truly their own:

> It is a process of empathy. Its origin is the indolence of the heart, that *acedia* which despairs of grasping and holding the genuine historical image as it briefly flashes up. Among medieval theologians *acedia* was regarded as the root cause of sadness. (391)[6]

---

[6] *Acedia* (also written *accidie*) played an important part in Benjamin's earlier work on the Baroque tragic drama. It names a state of listlessness, apathy, faltering ability to

Whether or not we find it convincing that the medieval idea of apathy and despair, *acedia*, should be the origin of such historiography, it follows that this abstraction of time into the empty, universal, homogeneous chronological medium of progress and continuity inevitably binds the future like a string of pearls trapped on a single, unchanging, unending thread that distracts us without leading from anywhere to anywhere.

Stalin's ticking clocks rest on a more systematic application of that same principle. He penetrates to the very core of what is most normal in military organization—every soldier has a watch and knows the time. A modern army articulates time with the utmost precision and discipline in order to maintain its coherence and its ability to deploy and maneuver. In the same way, every airline runs according to a precise timetable. But the signal conveyed by Stalin in the empty ticking renders all this as nothing. He makes time impenetrable because suddenly the besieged force looks back on a moment that broke against them, and sees an image of future time that will break again.

If it is correct to state that the formula linking time and terror operates both in the power to inflict and the power to resist, then we naturally ask: where do we find a master of terror who masters time in the opposite role? What are the resources by which we resist terror and in what opposing manner of time would we discover that mastery? In Poe we see how an acknowledged master transfigures terror by exploring the powers of narrative to penetrate what at first seem to be terrifying obscurities of time. Not all his works explore terror, but they all, in the most astounding variety of ways, explore obscurities of time. Indeed, his tight focus on the development of narrative time produces a supremely penetrable narrative manner.

One notices at once how Poe uses the motif of clocks and similar devices measuring out time in a mechanical way, but the full significance of this emerges when we remember that all narrative in any tradition constructs time according to its own characteristic sense of a vulnerability in the present. Some event, usually a misguided action or the violation of a norm, initiates the span of narrative time in a story by interrupting what had continued up to that moment as a sequence capable of indefinite

---

decide, failing ability to act, or to feel the value of choosing one thing over another. Among the literary figures in whom he sees this state exemplified, we find, as one would expect, Hamlet. The Church by tradition saw the symptoms of *acedia* in an inability to work or to pray. It is derived via the Latin *acedĭa*, from the Greek ἀκηδία, negligence, whose root terms mean "without care."

projection into the future. This is what endows the story with a shape originating at a beginning, passing through a middle, and arriving at an end. More explicitly, the story unfolds as the consequence of that break and fulfils its course until closure is reached at the point where narrative logic reinstitutes the regimen of normal time. This reinstated condition is not "normal" or normative in that it returns us to a concept of "truth or duty." It may well include such values for the protagonist and the cultural milieu in which it sets an action, but we as readers recognize them in another function. We feel the sense of a fictional closure when we see that the narrative has imagined a condition of time restored to permanence and continuity into the future.

Poe identifies the qualities of time in which we feel at home because they appear transparent and consistent, and he creates events that introduce those in which we feel alien, adrift, oppressed, because they harbor blind places and hidden shocks. The failure to understand the human present proceeds from the desire elicited by progress to identify with a universe that operates as a clock does, obeying the laws of nature according to cause and effect. The desire is provoked, on the one hand, by the apparent promise of causality to render all things according to a normal shape of reality, confirming the basis of our identity as homo sapiens, the human as knower. But this course of nature is also the curse of nature, bringing inevitable catastrophes or our inevitable end. Poe recognizes that this burning passion for knowing pitches us, on the other hand, all the more into the torments of vulnerability when the failure of knowledge robs us of our identity.

The law of terror in his world determines that the greater our power to reason, the more relentlessly we shall uncover the limitations and the failures of reason. His tales of terror illustrate this when a highly rational protagonist encounters a phenomenon at the limits of his comprehension, as we noted in "The Man of the Crowd." Sometimes Poe conducts the reader into the interiority of the insane or criminal mind of an individual who is nonetheless able to articulate his reflections and his delusion with all the care and transparency of reason. Examples of this include "Ligeia," "The Tell-Tale Heart," and "The Imp of the Perverse." Frequently Poe makes it difficult to decide exactly where the narrator's delusions begin and end, but he always illustrates the compulsion to speak, and the need of the human mind to explain itself. Above all, Poe explores the horror of forces that control his protagonist from places alien to that person's own conscious purposes and reason. The terror that Poe transforms into an aesthetic effect may be exterior in origin, or internal to the mind itself, and often the uncertainty of distinguishing between the two produces his most elevated effects.

The introductory remarks by the narrator of "The Black Cat" illustrate this pattern particularly well because the words he ostensibly pens in his last hours call on some unknown reader to provide the explication he craves in this world or beyond. The author fashions his story by letting the reader see through the language in which the narrator recounts his failure to see. Poe thus fulfills the magisterial role of calm and rational assessor of the events that the language itself describes as absent.

The text limits itself solely to the confession of a prisoner awaiting execution for the murder of his wife. According to the subsequent account, the arrest had followed from his failure to maintain his silence at the place where he had successfully concealed the body. By his own confession, he had split her head open with an ax when she tried to keep him from striking a black cat that he suspected was manipulating and persecuting him by supernatural power. Writing in his cell, he explains the purpose of this account:

> [T]omorrow I die, and today I would unburden my soul. My immediate purpose is to put before the world plainly, succinctly, and without comment a series of mere household events. In their consequences they have terrified—have tortured—have destroyed me. Yet I will not attempt to expound them.... Hereafter, perhaps some intellect may be found which will reduce my phantasm to the common-place—some intellect more calm, more logical, and far less excitable than my own, which will perceive in the circumstances I detail with awe, nothing more than an ordinary succession of very natural causes and effects. (349)

One can say that what he desires more than anything else—more even than a future life—is to find a way to frame his experiences within an ordinary flow of events. The succession of causes and effects here constitutes a definition of normal time. Poe does not fulfill his character's desire, but he does fulfill the complementary desire of the reader to enter into a completed course of narrative time. The story does not solve the narrator's implicit question either—we are not given decisive evidence within the narrative how to incorporate the possibility of witchcraft. What we have instead is our pleasure in a story that rises above the torment it portrays because the calm, logical, unexcitable mind that frames it has mastered the course of narrative time with a unique degree of precision. Though Poe in his personal life may have been as inclined to excitability as his characters, as a writer he personified the most perfect discipline. Baudelaire notes with approval that Poe constructs the mechanism of his narratives with the accuracy of an intricate machine. But while the sound of a ticking clock at the Battle of Stalingrad may heighten the terror of waiting, what Poe singles out as the essential purpose of narrative heightens the constructed coherence of a fiction in contrast with such horror. The difference lies in the order of time that links each articulation in the body of the tale to its dénouement. All well framed works of art offer aesthetic time. Narrative always makes the promise of a closure, but Poe's absolute aestheticism, spurning the didactic elements of "truth and duty" that link back to life, permits him an absolutely complementary order of unfolding time. In life, it is likely to be the power of disruption that puts an end to the hope for order. Time runs out, and a disaster ensues for which we are never sufficiently prepared. The work of art ends by closing the frame of its composition, a closure that unfolds in perfect consistency from its opening. And in that way, aesthetics masters terror by mastering the unfolding of time.

With its long history and the repeated disasters it has endured, Old Europe, or certainly a large part of its literary modernity, has found a necessary realism in this opposition between aesthetics and the worldly order. America, always eager to build a shining city on a hill, faithful to taking care of business in a world that needs cleaning up, has recoiled from its brilliant son Poe. Cantankerous critic Harold Bloom ranks Poe as tied for twelfth place among nineteenth-century American poets (3). T.S. Eliot, in an essay entitled "From Poe to Valéry," tries a truly odd series of postulations to account for the influence of Poe on the modern French poets, none of which concedes that they might have found something of real value in him. In that essay, Eliot writes that Poe has the mind of a precocious pre-pubescent boy. He lacks, Eliot explains, that which distinguishes "the mature man" which is, in his view, "a consistent view of life" (35).

Adrift in a world fraught with tensions and obscurities, we need to learn to derive an alternative source for a steadiness in our vision from an origin that lies outside the realm of natural processes. Benjamin seems to miss what he himself invokes as the "insights of Poe" when he tries to lead from that origin in a modern intensification of literary art to an anti-aestheticism in the twentieth-century avant-garde. But Eliot arrives at an equally misguided rejection of Poe's aesthetics by returning to the values of a traditional world order invoked as the views of a mature man who accepts and in some didactic spirit propagates a conventional ideology. We live amid a cacophony of ticking time bombs; we struggle against an impenetrable thicket of threats to our planet, our social order, our financial structure. Eliot, who famously described himself as "a classicist in literature, a royalist in politics, and an Anglo-Catholic" clearly considered a "consistent view of life" to mean living in hopes that whoever controls the forces modern humanity has devised may remain "more calm, more logical, far less excitable" than those who inflict such terror upon us ("To Criticize the Critic" 15). I would suggest we are more mature when we remain calm in the knowledge that sometimes they do not.

The lesson we can learn from Europe's appreciation of Poe shows how language, preserved in its integrity under the gentle spell of the muses, preserves us from irrational panic that rushes toward worldly decisions in what will, all too often, be an untimely manner, struggling, in Rumsfeld's eternal phrase, to manage "unknown unknowns." We see much to regret in the way Old Europe has responded to changes in its population, and much to acknowledge in the way that the United States has accepted its new citizens. Yet there is also a subtle dimension of difference that we see in the kind of language in which each realm conducts its discussions of policy and tradition. That has framed quite different discourses on the right to wage war, to detain suspects without trial, and most especially, the right to violate bodies by torture. A different notion of tradition in Europe, and a different hold on its origins, seem to have produced a different steadiness in its hold on the future. That tradition has been through terrors before, and has come back to its sense of itself. We can hope that the United States, too, might come back to the sense of its originating values, and preserve the dignity of its Constitution even in moments of the dire threat by which violence has insinuated itself into the fabric of time.

References Cited

Aristotle. *Poetics I.* Trans. Richard Janko. Indianapolis and Cambridge: Hacket Publishing Company, 1987.

Beever, Anthony. *Stalingrad.* London and New York: Viking, 1998.

Benjamin, Walter. "The Author as Producer." Trans. Edmund Jephcott. *Selected Writings 2.* Ed. Howard Eiland and Gary Smith. Cambridge, Massachusetts and London, England: The Belknap Press of Harvard UP, 1999. 768–782.

——. "Surrealism: The last Snapshot of the European Intelligentsia." Trans. Edmund Jephcott. *Selected Writings 2.* Ed. Howard Eiland and Gary Smith. Cambridge, Massachusetts and London, England: The Belknap Press of Harvard UP, 1999. 207–221.

——. "On the Concept of History." Trans. Harry Zohn and Edmund Jephcott. *Selected Writings 4.* Ed. Howard Eiland and Michael Jennings. Cambridge, Massachusetts and London, England: The Belknap Press of Harvard UP, 2003. 389–400.

——. "On Some Motifs in Baudelaire." Trans. Edmund Jephcott. *Selected Writings 4.* Ed. Howard Eiland and Gary Smith. Cambridge, Massachusetts and London, England: The Belknap Press of Harvard UP, 1999. 313–355.

——. "The Storyteller: Observations on the Works of Nicolai Leskov." Trans. Edmund Jephcott. *Selected Writings 3*. Ed. Howard Eiland and Gary Smith. Cambridge, Massachusetts and London, England: The Belknap Press of Harvard UP, 1999. 143–166.

——. "The Work of Art in the Age of Its Technological Reproducibility." Trans. Edmund Jephcott. *Selected Writings 3*. Ed. Howard Eiland and Gary Smith. Cambridge, Massachusetts and London, England: The Belknap Press of Harvard UP, 1999. 101–133.

Bloom, Harold, ed. *Edgar Allan Poe*. New York: Chelsea House, 1984.

Craig, William. *Enemy at the Gates: The Battle for Stalingrad*. New York: E.P. Dutton, 1973.

Eliot, T.S. "From Poe to Valéry." *To Criticize the Critic*. New York: Farrar, Strauss, & Giroux, 1965. 27–42.

——. "To Criticize the Critic." *To Criticize the Critic*. New York: Farrar, Strauss, & Giroux, 1965. 27–42. 11–26.

Kraus, Karl. "Der sterbende Mensch." *Worte in Versen* Vol. 1. Second Ed. Leipzig: Verlag der Schriften von Karl Kraus, 1919. 67.

——. *Last Days of Mankind: A tragedy in five acts*. Abridged and ed. by Frederick Ungar. Introd. by the editor. Critical analysis by Franz H. Mautner. New York: F. Ungar, 1974.

——. *Der Untergang der Welt durch die schwarze Magie. Werke 8*. München: Kösel-Verlag, 1960.

Lukács, Georg. *Theory of the Novel: A historico-philosophical essay on the forms of great epic literature*. Trans. Anna Bostock. London: Merlin Press, 1988.

Poe, Edgar Allan. "The Black Cat." *The Selected Writings of Edgar Allan Poe*. Ed. G.R. Thompson. New York and London: W.W. Norton & Co, 2004. 348–355.

——. "The Man of the Crowd." *The Selected Writings of Edgar Allan Poe*. Ed. G.R. Thompson. New York and London: W.W. Norton & Co, 2004. 232–239.

——. "The Philosophy of Composition." *The Selected Writings of Edgar Allan Poe*. Ed. G.R. Thompson. New York and London: W.W. Norton & Co, 2004. 675–683.

——. "The Poetic Principle." *The Selected Writings of Edgar Allan Poe*. Ed. G.R. Thompson. New York and London: W.W. Norton & Co, 2004. 698–704.

Shakespeare, William. "Hamlet," *The Globe Illustrated Shakespeare* Vol. 3. Ed. Howard Staunton. New York: Greenwich House, 1986. 1851–1934.

Solomon, Samuel. "L'espacement de la lecture: Althusser, Derrida, and the Theory of Reading." *Décalages* 1.2 (2012): 1–26. <http://scholar.oxy.edu/decalages/vol1/iss2/4>.

Zeller, Christoph. "Language of Immediacy: Authenticity as a Premise in Benjamin's *The Work of Art in the Age of Its Technological Reproducibility*." *Monatshefte* 104.1 (2012): 70–85.

THE FUTURE OF AUGUST 6TH 1945: A CASE OF THE 'PEACEFUL UTILIZATION' OF NUCLEAR ENERGY IN JAPAN

Masae Yuasa

Abstract: This article attempts to explain the enigma of Japan's nuclear policies, which I claim cannot be fully comprehended by mere economic calculations, political ambitions, or the bureaucratic interests of decision-makers. I use the concept of "time imagery" developed by Maki Yusuke[1] and argue that Japan's devastating defeat in 1945—marked by two atom bombs—became the origin in a segmented linear time imagery. In this imagery, the *telos* was constituted by the vision of a technotopia that was the anti-reality of 1945. This imagery became institutionalized immediately after the war, and it materialized in the steady comprehensive expansion of Japan's nuclear project. The first part of the article offers an overview of the enigma of Japan's nuclear energy policies. The second part explains the features of this enigma more closely, drawing specifically on Maki's theory of time imagery. In the conclusion, I evaluate Maki's model of four types of time imagery and also touch upon Japan's possible future nuclear policy after the Fukushima accident.

Keywords: Japan, nuclear policy, atomic bomb, time imagery, Fukushima accident.

Introduction

On March 11, 2011, a 9.0 magnitude earthquake and tsunami hit the Fukushima Daiichi Nuclear Power Plant. Both the pressure and containment vessels of the three nuclear reactors were destroyed in a meltdown of the reactors (Yoshioka 2011). No one yet knows the exact cause or mechanism of the accident, and there is speculation that it will take a decade or more to establish the precise causes because of the high level of radiation at the site. The Cesium 137 released in the air is estimated to be 168 times more than the Hiroshima atomic bomb (*Asahi shinbun* 2011a). But again,

[1] Throughout this paper, Japanese names will be given in the order customary in Japan, with the family name preceding the personal name.

no one knows the exact amount of Cesium, or even how much poured into the ocean. Before the plants could be shielded, radioactive materials continued to be released into the air, soil, and oceans.

According to a report on evacuees by the Reconstruction Agency (2012) established in the wake of the disaster, 62,000 people were evacuated from the Fukushima prefecture before the end of March 2012. Among those, about 25,000 cannot hope to return home for at least five years due to severe contamination (*Yomiuri online* 2012a). Contamination is, however, an issue that is not restricted to the Fukushima people. Not only can small radioactive particles travel anywhere in the globe through the wind and rain, but such particles are easily distributed throughout the economic market. For instance, green tea leaves produced in the Shizuoka prefecture, 360 km from the Fukushima plant, were severely contaminated by Cesium 137, and the tea shipment was subsequently banned (*Yomiuri online* 2011). As well, the government bought up 3,500 cows in seventeen prefectures that were fed with Cesium contaminated straw sold in the market (*Asahi shinbun* 2011b). Invisible radiation is ubiquitous and persistent. The impact of the accident on the society is enormous, but a radical transformation of Japan's nuclear policies is not evident.

In the spring of 2011, soon after the devastating tsunami, then Prime Minister Naoto Kan announced that he wanted to think openly about Japan's future energy policies. But Kan was replaced in office by Yoshihiko Noda in the autumn of 2011. Noda soon declared Japan's *re*commitment to developing 'safe' nuclear power to the UN General Assembly. His announcement was followed by news that the Japanese government had restarted negotiations with India on a controversial Nuclear Power Pact, and had reached an agreement with Vietnam on exporting its nuclear plant technology. This course of action is fully in line with Japan's previous national nuclear policies of the past 56 years.

Indeed, Japan's general nuclear policies have been both puzzling and irresponsible. While the country constantly refers to itself as "the first nation that was the victim of the atomic bombs," and has declared "no more Hiroshimas/Nagasakis," it has never officially questioned the legitimacy of nuclear weapons under the Japan-US Security Treaty. Furthermore, Japan has actively developed nuclear power technologies as part of its national nuclear policy since 1954. The country's firm commitment to the "peaceful utilization" of nuclear energy has never wavered, even after the 1986 Chernobyl catastrophe. The irony is especially acute in Hiroshima, a mere fifty miles from the town of Kaminoseki, where the Hiroshima-based Chugoku Electric Power Company (CEPC) proposed to construct a nuclear power plant. Despite fierce protests from local residents, most people in Hiroshima, who one would expect to be uniquely sensitive to the nuclear issue, remained silent, at least until the Fukushima accident.

This article approaches the paradox of Japan's nuclear energy policy in relation to the concept of time. Certainly, Japanese nuclear policy can also be explained as a business calculation.[2] Or we can speculate about the powerful ambitions of some of Japan's leading politicians.[3] Although these points are relevant, they do not tell the whole story. I will argue, based on models from the comparative sociology of time proposed by Maki Yusuke, that a time imagery shaped by concepts of a specific "*archē*" and "*telos*" captures postwar Japan's view of its future role on the world stage and in its own imaginative history. Reflection on these concepts helps to elucidate how, for Japan, nuclear power is the key to a specific kind of utopia. Maki's four types of time imagery and the specific case of the town of Kaminoseki will serve to interrogate Japan's problematic relationship with nuclear power in the post-Fukushima era.

Part I: *The Enigma of Japan's Nuclear Policies*

1. *Residents Move against Nuclear Power Plants: The Case of Kaminoseki*

Kaminoseki is a small fishing town in the Seto Inland Sea in Japan (see Figure 1). In 1982 the Hiroshima-based Chugoku Electric Power Company (CEPC), a regional monopolist, proposed the construction of a nuclear power station on the Tanoura seashore. Mountains isolate the place from even the nearest village; only Iwai Island faces Tanoura from across four kilometres of sea. In 1989 the Kaminoseki town council decided to accept the plan by a majority vote. However, ninety percent of Iwai's people rejected it.

The people on Iwai Island were convinced that the plant would destroy the natural resources on which their lives depend. Moreover, there is no place for them to hide and no way to escape, should an accident happen.

---

[2] According to the International Atomic Energy Agency (2009, 1–2), as of January 1, 2010, there were 437 nuclear power reactors in operation. Fifty-five reactors were under construction, the largest number since 1992. Of the eleven already under construction, ten are in Asia, as are thirty-six of the fifty-five reactors under construction. Furthermore, more than sixty countries have informed the Agency that they might be interested in launching nuclear power programs. In June 2009, the Japanese government set up the International Atomic Cooperation Council to facilitate cooperation among government, industry, and academia in order to export Japanese expertise on nuclear technologies (Katori 2009).

[3] There have been several politicians who are interested in nuclear armament (Nakashima 2004, 62–63).

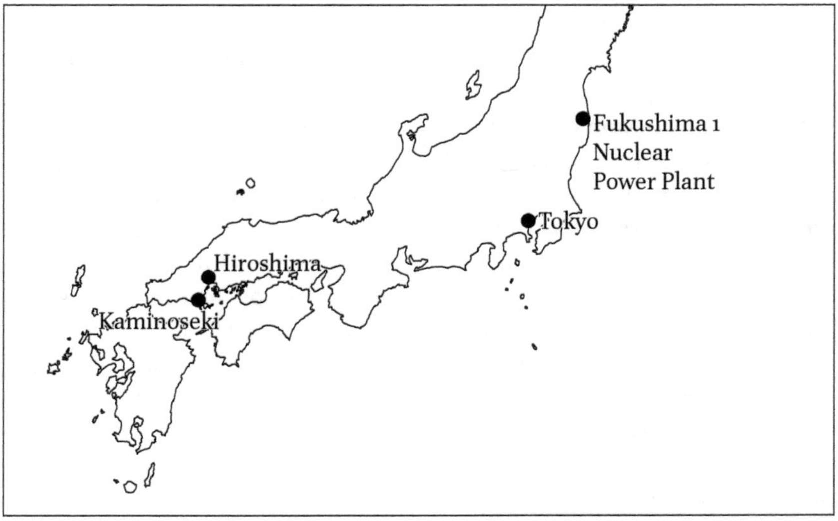

Fig. 1. Location of Kaminoseki.

They believe that the nuclear power plant might eventually desolate the island and end more than a thousand years of eminent history.[4] In 1994, the government designated Kaminoseki as an "important source of electricity" (*yōtaisaku jūyō dengen*), thus endowing the plan with legitimacy as a state policy (Asahi shinbun Yamaguchi Branch 2001, 106). Supported by this authority, the CEPC started the proceedings set by the government to build the plant. In turn, the inhabitants of Iwai responded in a variety of ways.

They staged direct actions in various locations: on their own island, in the Kaminoseki town council, in Tanoura, in front of the prefecture government building, and at the gate of the CEPC. Indeed, since September 2009, they have continued their sit-in blockade on the Tanoura seashore every weekday until the Fukushima accident, in order to stop the landfill work that the company was trying to commence. Besides these direct actions, they have also fought several court cases and elections. The protests became part of their lives. However, these actions were also demanding and burdensome because most of them were already over seventy years old.

[4] The name of the island appears in *Manyōshū*, the oldest poetry book in Japan, edited around the seventh to eighth century. Their festival, *Kan-mai*, has more than a 1000-year history.

Even before the Fukushima accident, the Iwai residents were not alone in opposing the plant. Three Japanese academic societies specializing in biology—the Ecological Society of Japan, the Ornithological Society of Japan, and the Japanese Association of Benthology—sent ten open letters to the government and the CEPC, claiming the company's environment assessments underestimated the damage to rare and endangered species living there (Sato 2010). As well, geologists criticized the company's insufficient research that failed to recognize the danger of fifteen active fault lines within 30 km of the planned site, including two lying precisely beneath the plant (Ochi 2010). In October 2009, 610,000 petitions from all over Japan were sent to the Ministry of Economic, Trade and Industry (METI), asking the government to reconsider the designation. However, all these claims failed to stop the plan. The company finally submitted an application for construction of an atomic reactor to the METI in December 2009.

Kaminoseki's enduring protest was not an exception. Since the 1970s, the government and electric power companies had difficulties acquiring new sites. In 2010, Japan had fifty-four nuclear plants in operation (*Citizen's Nuclear Yearbook* 2010, 58). However, most of the sites were proposed before 1970. Since then, only one locality has accepted the building of a nuclear plant, while thirteen either successfully rejected such plans or are still fighting them, among them the people from Iwai (*Citizen's Nuclear Yearbook* 2010, 59).

2. *Promoting Nuclear Power Plants as National Policy*

In his detailed study *Social History of Nuclear Power*, the historian Yoshioka Hiroshi examined Japan's nuclear energy policy of the last fifty years. He characterized its decision-making structure as a "binary sub-government model" (1999, 20). In this model two groups have dominated the decision-making process, while fighting each other over competences. One group is an alliance between the METI, the nine regional monopolistic electric power companies, government controlled banks, and the METI controlled Electric Power Development Company. The other is a Science and Technology Agency[5] (STA) group, which includes STA's two affiliated corporations and its national research centers. The former is in charge of the business stage and the latter of the technological development.

[5] STA was integrated into the Ministry of Culture, which in 2001 became the Ministry of Education, Culture, Sports, Science and Technology (MEXT).

No outsiders, including even the Japanese Parliament, have ever had any say in their decisions; and the local citizens, who have the most vested interests, were asked to "understand" or "agree with" the national policies (Yoshioka 1999, 27, 290).[6]

These two groups have overseen the remarkably "steady comprehensive expansion" of Japanese nuclear energy projects (Yoshioka 1999, 291). The term "steady" here signifies that both STA's and METI's budget and its installed capacity of nuclear power plants have been increasing (Figure 2). The term "comprehensive" means that development has covered almost all areas of civilian use of nuclear energy. The problem remains, however, that this expansion cannot be explained rationally in terms of public interests. The government's nominal reasons for its support have changed over the years. From the 1950s to the first half of the 1990s, it was "economics"; as well, from the last half of the 1960s the requirement for a "stable supply [of] energy" was evoked; and since the 1980s, the justification for nuclear energy was that it was "clean." However, each reason is based on rather one-sided judgments without a more contextualized and general serious examination (Yoshioka 1999, 294).

The dramatic events during the 1970s, such as the Nixon shock or the oil shocks, did not affect the said steady expansion. Fatal events such as The Three Mile Island accident in 1979 and Chernobyl in 1986 had almost no impact or influence. The government emphasized the uniqueness of those accidents and insisted that they could not happen in Japan (Nanasawa 1996; Yoshioka 1999, 217). A petition, signed by 3.5 million citizens, demanding a change in the policy also had little effect. Nevertheless, the rate of operation decreased due to various small accidents in the domestic nuclear plants: operation fell to 40–50% during the 1970s and was still less than 60% in the 2000s (International Energy Agency 2010).

The comprehensiveness of the expansion also seems difficult to justify. Japan has committed itself to the Nuclear Fuel Cycle project, once regarded as promising lasting energy abundance and autonomy. The key aspect of this project is the fast breeder reactor, which burns plutonium extracted from spent nuclear fuel. It supposedly produces more plutonium than

[6] This does not mean that the local governments or local people were reluctant to invite nuclear power plants. In his detailed study of the local community of the Fukushima Nuclear power plants, Kainuma Hiroshi (2011) argues that the local population desired to have the nuclear power plants in order to modernize their towns.

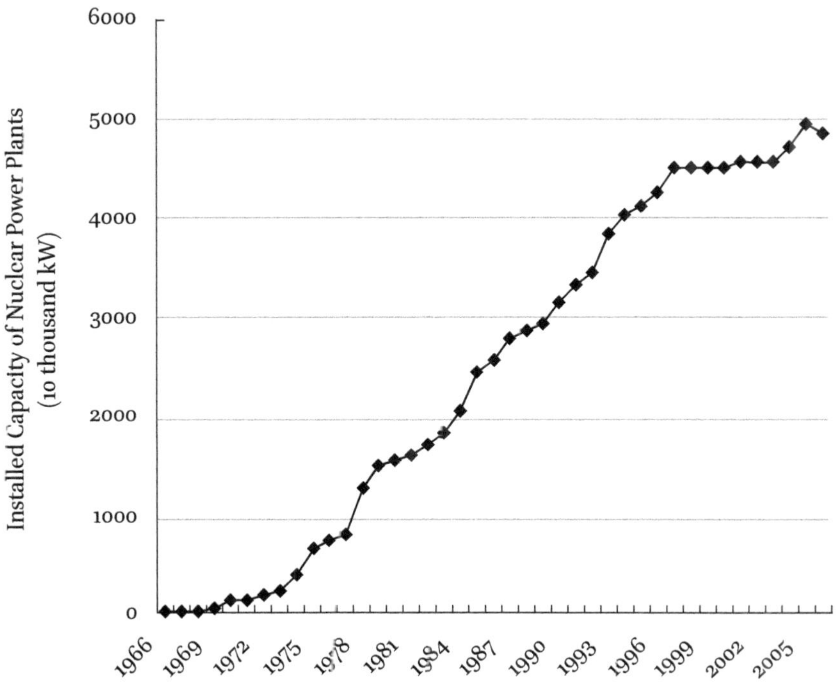

Fig. 2. Steady expansion of nuclear energy in Japan.
Note: the data come from *Citizen's Nuclear Yearbook* (Citizen's Nuclear Information Centre 2008, 310).

it uses, while at the same time generating energy. Many countries have already abandoned this project because it ran into a technological deadlock and produced huge operational costs.

The only Japanese fast breeder reactor, *Monju*, had a sodium leak problem and stopped operating in 1995. This could mean the end of the project since seventeen years are long enough to lose the original architect, engineers, and experienced operators who are vital for its safe operation.[7] In terms of economics, however, it has continued to consume fifty-five million yen every day in maintenance costs. As a result, the cost of the building and maintenance has already exceeded two times more than that of an ordinary nuclear plant, although it can generate only one third

[7] This comment was made by Dr. Imanaka Tetsuji, a researcher at the Kyoto University Research Reactor Institute, in his lecture on 25 April in 2010 at We Plaza in Hiroshima.

of the power of a standard plant (Okano and Nishiyama 2010). Despite all this, the government described the Nuclear Fuel Cycle project as a "foundation for the national nuclear policies" (METI 2010). The project thus seems beyond cost and risk calculation.

Based on his collected data, Yoshioka concluded that there was no evidence that Japan's nuclear energy policies over the past fifty years were based on prudent and sound political judgment (1999, 295). The only explanation he could find were bureaucratic interests of the decision-makers themselves; in his view, they pursued policies to maintain and expand their own group assets, such as power, budget, and employment (1999, 290–293). But this self-interest is a general tendency in bureaucratic organizations in any policy area, and Yoshioka himself is well aware that bureaucracy alone cannot explain everything. Thus, he suggests that "the need for the nuclear energy project" might be the axiomatic "central belief" of the cognitive paradigm among the decision-makers (1999, 305). This paradigm prevents them from critically questioning whether the promotion of nuclear power is appropriate for Japan. But how has "the need for the nuclear energy project" become the central belief of the paradigm? How was the paradigm born?

I wish to argue that the central belief was born when two atomic bombs were dropped in Hiroshima and Nagasaki in 1945. Maurice Merleau-Ponty's concept of "institutionalization" seems to describe the process. For him "institutionalization" is an event that gives enduring dimensions to a certain experience:

> What we understand by the concept of institution are those events in experience which endow it with durable dimensions, in relation to which a whole series of other experiences will acquire meaning, will form an intelligible series or a history—or again those events which sediment in [us] a meaning, not just as survivals or residues, but as the invitation to a sequel, the necessity of a future. (1970, 40–41)

Based on this concept, Kazashi Nobuo (2010a) has argued that dropping the atomic bombs in Hiroshima and Nagasaki is the origin of institutions of the post-Hiroshima world order. Although he discusses this at the global level, the same logic seems applicable to the national level. In what follows, I will draw on the Japanese sociologist Maki Yusuke's theory of "time imagery." Maki's theory elaborates Merleau-Ponty's poetic phrases and powerfully explains the enigma of why the country to fall victim to the atomic bombs is intent on pursuing such an active policy of nuclear energy use.

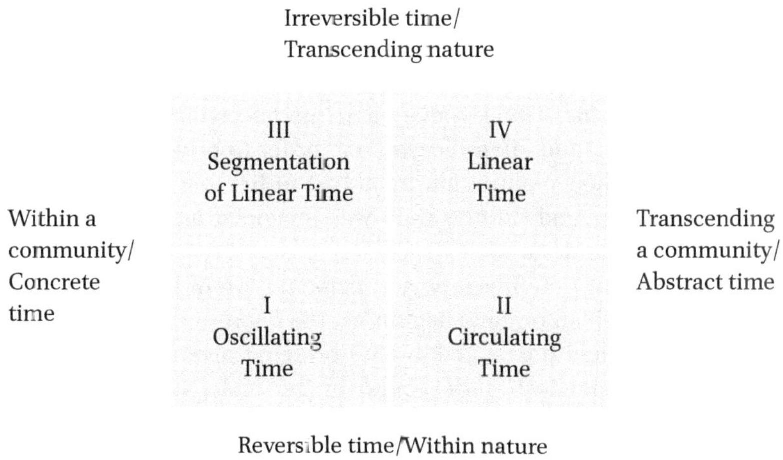

Fig. 3. Four ideal types of time imageries.

Part II: *Explaining the Enigma*

1. *Time Imageries and Institutionalization*

Let me first briefly introduce the theory of time imagery: Maki (2003) proposes four ideal types of time imageries across cultures (see Figure 3). The horizontal axis shows the relation between time and community; the left is concrete and qualitative time inherent in a community and the right is abstract and quantitative time transcending beyond a community. The vertical axis shows the relation between nature and time; down is reversible time inherent in nature and up is the irreversible time in a human-made world transcending nature. Maki (2003, 312) also argues that a certain mode of culture or social structure serves as grounding for each time imagery.

The first type of time imagery is oscillating time, which is seen in most technologically primitive communities. Here time is inherent in both community and nature. Two different times exist in parallel. One of these is the time of Nature/God, which is eternal. This Nature/God time gives order to the other time in the community life by providing meaning to past events and predicting the future as the realization of Nature/God time. This Nature/God time represents potential, while the other community time is explicit, realizing the potential. Elsewhere, similar time imagery is found in the totem theory by Claude Lévi-Strauss, or the sacred time theory by Mircea Eliade (Maki 2003, ch. 1).

The second type of time imagery is circular time, based on the time inherent in nature as perceived by agricultural communities. This type of time imagery was embodied in, for instance, ancient Greek society. Maki (2003, 179) argues that Greek society and similar civilizations needed to develop an abstract and quantified time in order to bridge different communities within their realm. Time thus had to be objectified and calculated. This abstract and infinite time was imagined as a circle, a figure representing infinitude to the Greeks.

The third type of time imagery is a segmentation of linear time (Maki 2003, 183–196), with an original beginning, the *archē*, an (imperfect) present, and an imagined future end, a *telos* bringing history to completion. This is the eschatological model found in the Bible, such as in Daniel, Ezra, and the Apocalypse. We witness this time imagery in the prophets in whose words the Jewish people found hope in their suffering and despair. In a more general view, when a community suffers, people often imagine returning to happier days. When their pain reaches despair and hopelessness, they long for a state of perfection, a utopia purified of all evil, where their suffering will be redeemed. The future utopia, however, is grounded in the original *archē*. This utopian future can acquire a certain reality through the collective dreaming of the community.

The last type of time imagery is modern linear time (Maki 2003, ch. 4), which transcends both nature and community. Here time flows and never returns. This time is also linear, but it differs from Hebrew time, which is oriented with respect to an origin and a goal. Modern linear time is without such a *telos*. We are compelled to keep moving toward the future, but because of our limited individual lives, we eventually 'arrive' at an empty death.

In the following, I will argue postwar Japan was captured by a time imagery of the third type, and that this contributed to the formulation of Japan's nuclear energy policy.

2. *6th August 1945 as an Origin for Japanese Nuclear Policies*

For the Japanese, the atomic bomb attacks did not signify only the devastation of two cities, but also the end of the morass of war and the start of a new Japan. John Dower (1999, 494), in his book *Embracing Defeat* writes that "the terrible power of nuclear weaponry proved as mesmerizing as it was terrifying, for nothing better exemplifies America's superior scientific, technological and organizational capabilities."[8] To prove this, he explains how the atomic bombs acquired new meanings.

---

[8] Similar comments were made by Morris-Suzuki and Yoshimi (2010, 61).

The 15th of August, the day of Japan's surrender, Prime Minister Suzuki said in his speech that the biggest shortcoming of the war had been "science and technology." The next day, the Minister of Education said the nation's task was to elevate "science power and spiritual power" to the highest level. The *Asahi Newspaper* declared on the 20th of August, "We lost to the enemy's science": "This was made clear by a single atomic bomb dropped on Hiroshima." Another article entitled "Toward a Country Built on Science" emphasized that "science" had to be understood in its broadest sense as involving "reason" and "rationality" in all spheres and at all levels of society.[9] These words materialized into political institutions. On the 17th of August, the Ministry of Culture announced the formation of a new bureau of scientific education. The new minister of culture proclaimed that "the cultivation of scientific thinking ability" was the key to "the construction of a Japan of culture." The government publicized its plan to divert 500 million yen from former military funds to promote science in everyday life (all quotes from Dower 1999, 494).

Dower (1999, 494) goes on to say that "science" soon became almost everyone's favorite concept for explaining both why the war had been lost and where the future lay. For the wartime leaders, "science" was a self-serving and pragmatic justification for the cause of the defeat. However, the term was also welcomed by ordinary Japanese, who were overwhelmed by grief and guilt towards their own dead countrymen and women (486). A victory might have validated those deaths, but defeat compelled the justification of those sacrifices. "Science" would be the answer. Many understood the reason for the defeat as "irresponsible leaders operating out of inherently backward, irrational, and repressive institutional structures" (495). Thus, Dower concludes, both the consciousness of having been victimized and the "question of responsibility for defeat" led many Japanese toward a relentless commitment to a *rational* society (495).

A "Country built on Science" was not a new concept for post-war Japan. But the crucial point here is that the terms "science" and "technology"

---

[9] It should be noted that for the Japanese the terms "science" (*kagaku*) and "technology" (*gijutsu*) are difficult to separate; science usually means "scientific technology" (*kagaku gijutsu*). Sasaki Chikara (1996, 18) says this usage can be traced back to the Meiji era, when the term "science" was imported to Japan and translated into "*rigaku*," which meant both pure science and engineering. During this time, the Japanese Meiji government actively imported western technology based on science in order to modernize Japan. It was also a time when in western societies, the concepts of "science" and "technology" had merged for developing "technology based on science." Yoshioka (1999) speaks of a "country built on technology," but he does not really differentiate science and technology according to common usage in Japan; he also uses the term "science-technology" interchangeably with "technology."

gained a powerful new connotation related to the atomic bombs and the United States. Furthermore, the term was stretched and overburdened, not only into a practical tool for national reconstruction, but also in discussions of national security, national pride, and even acknowledgment of the dead. Yoshioka points out that scientists and policy makers in Japan's "nuclear community" successfully internalized the concept, and therefore needed to be emancipated from it in order to recognize public interests (1999, 302). But this would not be easy. According to Merleau-Ponty, the process of institutionalization causes people to forget how the new dimensions of a certain experience were constructed in contingent situations (Hirose 2010, 187). The connotation keeps working in the institutions without people recognizing it or remembering its origin.

3. *A New "Country built on Science" as Anti-reality of the Old*

The motto "A Country built on Science" designates the idea that the strength and prosperity of nation states depends on their capacity to develop cutting-edge science. Yoshioka (1987, 194) argues that, although Japanese scientists had supported this idea since the nineteenth century, the Japanese government only started to commit to it during World War II. This attempt ultimately failed, but it provided important lessons for the post-war version of a "Country Built on Science."

As a latecomer to industrialization, Japan had been forced by outside pressure to transform quickly from a feudal into a modern nation. Without much time to develop its own innovative intellectual and social structure, it actively imported advanced Western technology as a national policy (Morris-Suzuki 1994, 73). After embarking on a policy of military aggression on the East Asian continent in the 1930s, and especially after opening war with the United States in 1941, increased international isolation forced the government to develop emergency measures to ensure its technological independence. Strategically, Japan could not rely on the enemy's technology, and from a purely practical perspective, Western technology had limited use because Western trade embargos caused a serious shortage of raw materials (Morris-Suzuki 1994, 144).

The Japanese Planning Agency, empowered by the National Mobilization Law of 1938, had issued ordinances on research in technological areas of key national importance. In 1941, it published the *Outline Plan for the Founding of a New Order for Science and Technology*, calling for the "establishment of a 'unique Japanese style' of science and technology based on the autonomous resources of the Greater East Co-Prosperity

Sphere" (Morris-Suzuki 1994, 148). A Technology Agency was created and expected to control and co-ordinate various policies under one roof. However, this newly established agency failed to incorporate technological divisions under other powerful ministries such as the Ministry of Education, the Ministry of Commerce and Industry, and the Army and Navy. Still, increased state budgets and political commitments towards technological development led to a sharp rise in the number of private and state-controlled research establishments. Nevertheless, the state ultimately failed to co-ordinate and mobilize them effectively for its war effort (Dower 1993, ch. 3; Morris-Suzuki 1994, 147).

By the end of the war many Japanese were convinced that the attempt to develop a "uniquely Japanese style of technology" had failed and that "the war of science and technology" had been lost (Morris-Suzuki, 1994, 157). This failure, combined with the atomic bomb experiences, marked the origin of a new vision of Japan in contradistinction to the old, exceptionalist view: Japan now was imagined as a nation to be built on universalistic science (as opposed to the "uniquely Japanese" science of wartime ideology) working within a framework of international co-operation. This vision was concretized into that of a "Country built on Nuclear Power," operating within the US nuclear regime—the perfect anti-image to the nation that had lost the war.

4. *Institutionalizing a "Country built on Nuclear Power"*

Yoshioka (1999, 55–88) describes how this vision was institutionalized within just three years. Under the Allied occupation, nuclear research was prohibited in Japan, and even after Japan regained independence in 1952, Japanese scientists did not collectively act to restart the nuclear research. There was strong opposition in the scientific community due to fears of the military use of nuclear power in the Cold War. However, after President Eisenhower delivered his "Atoms for Peace" speech at the United Nations General Assembly in 1953 and the new Atomic Energy Act allowed the US government to provide nuclear materials and technologies to its allies through bilateral agreements, Japanese politicians who were familiar with these radical changes quickly took action and prepared the first nuclear budget, which was presented to the Diet on March 2, 1954. The plan set aside 3,000 million yen to promote science and technology, of which 2,600 million were designated for building a nuclear reactor and searching for uranium. Once the budget was approved in the Diet, institutionalization promptly started. The government set up the Preparation and Investigation Committee for Utilizing Nuclear Power in May 1954. It worked as the supreme deliberative organ, deciding nuclear policies until the Atomic Energy Commission of Japan was established in 1956. The importance of the committee was clear in the lineup of members, including the deputy Prime Minster, the chief of the Economic Planning Agency, Minister of Finance, Minister of Culture, Minister of Economy, the chairman of the Japan Federation of Economic Organizations, and the chairman of the Science Council of Japan. The committee decided to sign the Japan-US agreement on nuclear research, and to accept concentrated uranium from the United States. The government also formed the Nuclear Budget Discussion Group within the Ministry of International Trade and Industry (MITI), which also formed an Overseas Nuclear Investigation Team. Based on its report, the group submitted a "Middle Term Plan" for building research reactors in 1955.

Immediately after the nuclear budget was tabled, the Science Council of Japan, an organization representing Japan's scientists, responded by making a declaration in April, 1954 of three principles for future nuclear research: "The Three Nuclear Principles" (*genshiryoku sangensoku*), i.e. "openness, democracy, and autonomy" (*kōkai, minshu, jishu*). However, these principles neither challenged the decisions on nuclear development nor did they become practical policy guidelines (Yoshioka 1999, 73). And in any case, the scientists soon jumped on the bandwagon after the budget was approved.

The industry also responded quickly. The Nuclear Documents Investigation Group was formed in December 1954, and in the next year, the Council for Peaceful Uses of Nuclear Power was established. These two were merged with the Power Economy Research Institute to set up the Japan Atomic Industrial Forum in 1956. As well, by August 1956 Japan's heavy electric machinery firms had quickly formed five industrial groups for nuclear business, including the three largest industrial conglomerates. It was as though the whole industrial circle jumped into the nuclear business despite its uncertain future.

In August 1955, the multi-partisan Nuclear Research Parliament Delegation was formed and attended the UN sponsored International Conference on Peaceful Uses of Atomic Energy in Geneva. Immediately after returning to Japan, members of the delegation issued a joint statement at the airport and soon established the Joint Committee of Atomic Energy, consisting of three representatives from each of the main four political parties in the Diet. This committee quickly prepared most of the legislative drafts relating to nuclear issues. At the end of the year, it submitted three major nuclear laws: the Atomic Energy Fundamental Act, the Act for Establishment of the Atomic Energy Commission, and the Act for Establishing the Nuclear Power Bureau. In 1956, other laws relating to nuclear development passed the Diet. Thus, the main administrative organs and state research and development institutions were set up at a fast pace, and without any opposition to speak of.

A seminal administrative organ at the time was the Science and Technology Agency. The Joint Committee of Atomic Energy had demanded a comprehensive and powerful ministry organization, which was expected to oversee and centralize all state scientific activities. However, facing powerful opposition from ministries such as MITI, only a rather small-scale agency focusing on nuclear technology was established. As in the case of the Technological Agency during the war, bureaucratic sectionalism hindered implementation of a "Country built on Science." Furthermore, MITI and the powerful industrial groups took the initiative to import nuclear technologies and buy nuclear plants instead of developing them on a national basis. Importing technologies was quick and economical, and the strategy fit perfectly with the post-war vision of a nation built on universalistic (instead of parochial) science and technology, and operating within a framework of international co-operation.

Once the foundational institutions were all set up, the main actors, such as politicians, bureaucrats, industrialists, and scientists, attempted to take advantage of them by fostering their own interests, while also adapting them to changing national and international circumstances. The axiomatic belief in a "Japan built on Nuclear Power" as the concretization of a "Country built on Science" had been firmly established and had materialized into a web of social and administrative institutions. In this way, the path was laid out for the "steady comprehensive expansion" of Japanese nuclear energy projects. The time imagery with Hiroshima as the *archē* and the "Country built on Science" as *telos* was embedded in those institutions—its origins already forgotten.

5. *Declining Community Support for the Time Imagery*

Although the segmentation of linear time imagery with *archē* and *telos* has been dominant and institutionalized in Japanese society, it seems to have weakened noticeably since the 1990s. Part of the reason is the decline of the community that had envisioned through its time imagery Japan's technological nuclear utopia.

After the war, the Unites States preserved part of the existing power structure in order to stem the rise of socialist organizations and to prevent a state of anarchy. Specifically, the US spared Emperor Hirohito, whom many believed to be responsible for the war, from war trials. Although the Emperor was transformed from a "living god" to a "symbol of national unity," he continued to reproduce communities of power, providing the authority that legitimized society's power structure (Yasumaru 2001). However, the famous "marriage" photo of Hirohito and General MacArthur,[10] the Supreme Commander of General Headquarters of the Allied Forces, clearly illustrated that the "unity of the nation" required support from the United States (Figure 4).

After Emperor Hirohito died in 1989 and with the end of the Cold War, the United States functioned less and less as a model for Japanese society and as a supporter of the Japanese power structure. The community of bureaucrats, politicians, scientists and industrialists who mainly sustained the time imagery of a "Japan built on Nuclear Power" had been losing its central figures. Since the 1990s, the political, economic, and social institutions that had supported this community have given way under the pressures of globalization. Also, the idea of 'Hiroshima' as an "original moment" of post-war Japan has been receding in the collective memory, ceasing to function as an *archē* for the nation. Under these circumstances, the time imagery that had operated in the comprehensive expansion of nuclear energy may continue to lose ground. In fact, Japan has been gradually moving toward a more modern time imagery. Most people, for instance, do not work for their 'country.' Rather, they are desperately pursuing their individual security and prosperity in an environment characterized by collapsing communities and decaying nature. Ironically, this model of action, directed by individual self-preservation, results in further social insecurity and anxiety. This vicious circle is the very foundation of modern Japan. Surprisingly, in spite of the changing time imagery, a large part of the population continues to support the nuclear energy policy—but without believing in a "Country built on Science." The old institutions remain and manage to keep channeling energies toward the instituted *telos*, through the media of power and money.

---

[10] The photograph was taken on 27th September 1945 at the US Ambassador's residence in Japan when the Emperor Hirohito visited General MacArthur. This was shown in Japanese newspapers two days later.

Fig. 4. Douglas MacArthur and Emperor Hirohito.

6. *Time Imagery of Advocates and Protestors of a Nuclear Power Plant in Kaminoseki*

The main advocates for a nuclear power plant in Kaminoseki town are those who are in power, such as the town mayor, his local officers, and members of the town council. The Kaminoseki Commerce and Industry Association is also an important proponent (Yoshizaki 2001). These claim to support the town development by inviting the construction of a nuclear power plant, and they sometimes even reference the "national policy" they are following.[11] Declining and aging populations, along with shrinking fishing and agriculture industries, have made the town financially unsustainable. The mountainous geography further reduces options for development. In this situation, and without much help from the prefecture and the central government, the income brought by the nuclear power plant is attractive. From 1994 to 2009, the town received three billion yen from the CEPC as a "donation." Beyond that, state subsidies will be around forty-five billion yen for ten years from the start of the plant's operation (Kihara 2010, 87). The Kaminoseki Fishery Cooperative received 12.5 billion yen for compensation from the CEPC. Individual proponents have also received various presents and monetary rewards from the company, although amounts are difficult to calculate (Kihara 2010, 188–190). In a situation opposed both to nature and community, the plan's proponents, as well as the company, attempt to solve problems with money. Their developmentalism and mammonism are committed to a modern, linear time imagery. But it is also possible that some people still retain the segmented linear time imagery, believing that they serve a "national community" and work towards the *telos* of a "Country built on Nuclear Power."

The people on Iwai Island, on the other hand, are different: while they have also suffered from a declining and aging population, they have kept a deep relation with nature and community. They had no possibility of commuting to outside workplaces and most have had to rely on nature. Although making a living through work in the primary sector is not easy, they have developed new products such as organic food for the outside market, and new ways to circulate them, such as selling their seafood and agriculture products through the Internet. They have also attracted visiting anglers, teaching and helping them fish on their boats. Thus, they still rely heavily on nature. They are often heard to say, "we can live as long as nature is preserved." For the elderly, their self-grown vegetables and the seafood they catch are sufficient for their survival, along with their pension. They have respected and appreciated nature. Nature has ordered their individual and collective works. To them, nature is the primary origin of order in their lives. After coming back from her protest, one woman murmured, "The very basis for human beings is protecting nature, isn't it?"

---

[11] A representative opinion can be read in a comment made by the president of the Kaminoseki Town Management Council (Yoshizaki 2001, 286).

This is a clear principle for her and the people on the island. She cannot understand why the principle is not true for others, such as the workers from the CEPC.

Communities are strongly alive and functioning on the island. There are several houses where elderly people get together every night to have a chat while drinking tea. Without much help from the outside world, including local government, they have found their own mechanisms of mutual help without mediating money. The three children in the Iwai elementary school are like grandchildren for all the residents. People go to see school events and enjoy themselves. They join the school ceremonies with formal wear. They resemble a big extended family. Furthermore, the local inhabitants keep strong communal ties with those who have passed away. The infrastructure they rely on was not made by the government or by someone's paid labor; it was their grandparents and great grandparents who developed the fields they continue to cultivate. They believe that their ancestors protected the nature that still provides for them. In their minds, the past and their ancestors are actualized in the means of production, landscape, and nature. Although young people leave the island for jobs, they come back during the holidays with their children and expect to return home after their retirement. The elderly on the island harbour a strong sense of responsibility for keeping their own houses and nature for their children, grandchildren, as well as for their ancestors.[12]

This feeling of responsibility towards both ancestors and posterity exemplifies the time imagery of the Iwai Island people, which appears to be close to Maki's first type of "oscillating" time. They expect the future to be like the past or the present, a realization of potential time inherent in nature.[13] For them, the nuclear power station will destroy their nature

---

12 I have heard such comments from people in Iwai when I was staying on the island. The way of life of the Iwai people and their voices are well documented by Hanabusa Aya in her film, *Hōri no Shima* ("Holy Island") that opened in theatres in 2010.

13 It is interesting to see small conflicts among the islanders towards "the Iwai island 100 natural energy project", which was initiated and financed by outsiders (http://www.iwai100.jp/index.html). The islanders were divided on this project even though it seemed to rescue their movement, which had already reached a desperate point just before the Fukushima accident. One explanation is that those holding on to the oscillation model of time imagery would not easily subscribe to this type of venture business, which projects a linear timeline into the future. It would be interesting to determine whether the supporters of the project have a different time imagery.

and hence their lives.[14] In their time imagery, there is neither place nor need for a grounding in an *archē* and an orientation towards the *telos* of a human-made technotopia. The slogan "A Country built on Science" means nothing to them. They have no illusions about nuclear power stations. Some of them have worked in several plants operating in different prefectures and talk about their fear of radiation. Their time is deeply rooted in nature and in the community, which in turn offers them order, stability, and peace.

Life on Iwai Island has attracted young people—typically those who are not part of the regular labor market—who feel alienated by the modern life style and by a society that still lives through the postwar time imagery. They, too, reject the post-war vision of a collective *archē* and *telos*, as well as the aimless linearity that has replaced it. Some have already escaped, creating their own communities and choosing to live with nature. They express admiration for the stability and steadfastness of the island's inhabitants and are impressed by their natural way of life.[15] They feel the life on Iwai Island is ideal, and they have joined the protests against the power plant. They think the government's nuclear energy policy is simply irrational. However, it is not easy to know the time imagery in the minds of these young people. Some have begun to live on the island following the lifestyle of the islanders, and they may eventually come to share the oscillating time imagery of the Iwai people.

As of fall 2012, the situation in Iwai was still unresolved. In February 2011, three weeks before the Fukushima accident, the CEPC sent more than 600 security men to the Tanoura seashore at two o'clock in the morning in an attempt to start landfill work without the consent of the islanders. This inevitably brought about a direct confrontation between the workers and the protestors. The tense confrontation continued for three days and was stopped only when two protestors were injured. At that moment, it seemed just a matter of time before the CEPC would

---

[14] They worry not only about the possibility of serious accidents, but also about the destruction of their fishing ground by shoreline stabilization, and emissions of vast amounts of hot water and chemicals into the sea. Also, they will lose the market for their high-value-added organic products due to rumors of even small amounts of radiation. Finally, they do not want to live in sight of the plant. They fear that if the plant is built, they will not be able to continue to live, that their children will not return, and that the island will become a no-man's land.

[15] Comments to that extent were often made by young people I met on the island, but the most eloquent testimony is the film *Hōri no Shima* ("Holy Island"), which vividly and beautifully represents director Hanabusa's love and admiration for the lives of the people in Iwai.

re-open their landfill works, employing coercive measures if necessary. The Fukushima accident has radically changed the tide. It is now less likely that the government will give the CEPC permission to build a plant in the green field site. On October 5, 2012 the governor of Yamaguchi prefecture refused to renew the CEPC's landfill license. Nevertheless, the CEPC refuses to abandon the plan in a situation they continue to evaluate as fluid (*Yomiuri online* 2012b).

Conclusion: Maki's Four Types of Time Imagery and Changes in the Time Imageries after the Fukushima Accident

This article has attempted to explain one seminal aspect of the enigma of Japan's nuclear policies, which cannot be fully comprehended only by taking into account economic calculations, political ambitions, or the bureaucratic interests of decision-makers. I have employed the concept of "time imagery" as developed by Maki and argued that Japan's devastating defeat in 1945, marked by two atom bombs, instituted the beginning of a segmented linear time imagery. The *telos* in this imagery is a techno-topian Japan representing the anti-reality of 1945. This vision redefined the century-old objective of creating a "Country built on Science." In the post-war version, 'science' came metonymically to indicate both nuclear technology and the United States. Science stood for the tools for rebuilding the country, but also for pride, security, and even consolation for the dead. This imagery was instituted immediately after the war, and was firmly embedded in the social institutions by the second half of the 1950s. Based on those institutions, Japan's nuclear policies realized a "steady and comprehensive expansion" of the nuclear project. Once the institutions were set up, they could be sustained and reproduced within the framework of an aimless linear time that replaced the post-war segmented time imagery.

However, it seems that the protesters fighting against state policies have escaped from the segmented postwar time imagery—if they were ever part of it. I have argued that the Iwai residents inhabit an oscillating time imagery, where time is embedded in the periodically recurring rhythms of nature and community. Their way of life, compatible with their image of time, successfully also attracts young people frustrated with the modern life style in Japan.

In this way, Maki's typology of time imageries not only helps us to understand the enigma of the Japan's nuclear policies, but it also elucidates the different positions of the actors in relation to the Kaminoseki issue and the nature of their 30-year struggle. Maki's comprehensive theory bridges individual psychology and social structure, and offers great potential for analyzing the dynamics of society. What, then, can we extrapolate concerning Japan's nuclear policy after the Fukushima accident? Will the enigma continue?

The Fukushima accident had an enormous impact on the social structure and has possibly unsettled the dominant time imagery. The vast amounts of radiation released may propel Japanese society into the modern linear time imagery in two ways. Firstly, life will be further removed from nature, as people fear the nature that carries, accumulates, and concentrates radiation. Secondly, more people are moving away from communal time after losing local communities when forced (by government policies or fear of radiation) to evacuate. Also, the fear of invisible radiation harms trust and deteriorates social ties within communities and families. Those factors push people to embrace modern time imagery.

As I have argued above, the modern linear time imagery supported and sustained the existing nuclear policy, even without consciously believing in its *telos*. A more powerful modern time imagery could work towards the continuation of the same enigma.[16] However, mounting distrust of the government and existing institutions after the accident has also pushed people to demand revisions to the nuclear policy, as they feel their own security and interests threatened. The question is whether without a new vision or a wide communal consensus on national objectives—and therefore, a new version of segmented time imagery—it will be possible to construct solid institutions to successfully transform energy policy into a more sustainable force. If the modern, linear time imagery remains dominant, policies might become more ad hoc, catering to the circumstances of the moment. Only the released radioactive materials will tick down to eternity.

[16] There is also a sign that people in power are attempting to make use of the current Fukushima disaster as another "*archē*" and renewing its commitment to "*telos*" by redefining the "Country built on Science." See for example an article by Yakushiji Taizō (2011), a member of the Cabinet Office's Council for Science and Technology Policy. The government seems to commit and promote scientific technologies in areas of decontamination, to the security of nuclear plants, and to the health management of people exposed to radiation.

References Cited

Asahi shinbun. 2011a. "Genpatsu jiko no hōshutsu seshiumu genbaku no 168bai: Hoan'in kōhyō" [Released Cesium from the Nuclear Power Plant Accident is 168 times more than that from an Atomic Bomb, the Nuclear Safety Agency Reported]. 26 August 2011.

——. 2011b. "Osenwara tabeta ushi subete kaitori Nōsuishō 3500tō" [The Ministry of Agriculture, Forestry and Fisheries Bought out all 3500 Cows fed with Contaminated Straw]. 5 August 2011 http://www.asahi.com/national/update/ (accessed 30 March 2012).

Asahi shinbun Yamaguchi Branch. 2001. *Kokusaku no yukue* [Traces of National Policies]. Kagoshima: Nanpō Shinsha.

Citizen's Nuclear Information Center. 2010. *Shimin genshiryoku nenkan* [Citizen's Nuclear Yearbook], Tokyo: Nanatsumori Shokan.

Dower, John. 1993. *Japan in War & Peace: Selected Essays*, New York: New York Press.

——. 1999. *Embracing Defeat*, New York: Norton.

Hirose, Kōji. 2010. "Shobunka wo ōdansuru setōteki na shinri" [Militant Truth Cutting Across Cultures]. *Gendai Shisō* 38(7): 186–197.

International Atomic Energy Agency. 2009. International Atomic Energy Agency Annual Report 2009: Vienna: IAEA.

International Energy Agency. 2010. "Energy Technology Perspective 2010." http://www.iea.org/speech/2010/Tanaka/NEDO_6Aug.pdf. (accessed 1 September 2011).

Kainuma, Hiroshi. 2011. *Fukushima-ron* [On Fukushima]. Tokyo: Seidosha.

Katori, Keisuke. 2009. "Kaigai genpatsu jinzai ikusei de kyōryoku" [Helping Nuclear Plants Overseas by developing Human Resources]. *Asahi shinbun*, July 10.

Kazashi, Nobuo. 2010. "Hibaku shintai to pawā/kenryoku: posuto-Hiroshima jidai no 'mierumono to mienaimono'" [Radiation-Exposed Body/Power: 'The Visible and the Invisible' in the Age of the Post-Hiroshima Era]. *Gendai Shisō* 38 (7): 174–185.

Kihara, Shōji. 2010. *Genpatsu sukyandaru* [Scandals of Nuclear Power Plants]. Tokyo: Nanatsumori.

Maki, Yusuke. 2003. *Jikan no hikaku shakaigaku* [Comparative Sociology of Time]. Tokyo: Iwanami.

Merleau-Ponty, Maurice. 1970. *Themes from the Lectures at the College de France*. Evanston: Northwestern University Press.

Ministry of Economy, Trade and Industry (METI). 2010. "The Fourth Consultative Meeting on Nuclear Fuel Cycle and Messages about Pluthermal Project." http://www.meti.go.jp/kohosys/press/0001570/0/0523kaku.htm (accessed 1 September 2011).

Morris-Suzuki, Tessa. 1994. *The Technological Transformation of Japan: From the Seventeenth to the Twenty-first Century*. Cambridge: Cambridge University Press.

Morris-Suzuki, Tessa and Yoshimi, Shunya. 2010. *Tennō to America* [The Emperor and the US]. Tokyo: Shueisha.

Nakashima, Shingo. 2004. "'Dōmeikoku Nihon'-zō no tenkan" [Changing images of 'Allied Japan']. In *Ikeda/Satō seikenki no Nihon gaikō* [Japanese Foreign Policies during Ikeda and Satou Governments], ed. Sumio Hatano, 53–92. Kyoto: Minerva.

Nanasawa, Kiyoshi. 1996. *Genpatsu jiko wo tou* [Questioning the Accidents at Nuclear Power Plants]. Tokyo: Iwanami.

Ochi, Shuji. 2010. "Kaminoseki chiiki no chishitu to jishin katsudō" [Geology of Kaminoseki Area and Earthquake Activities]. Paper presented at a meeting entitled, "Reading License Application for Constructing Atomic Reactor" at the International Conference Center Hiroshima, Hiroshima, September 23.

Okano, Shō and Nishiyama, Takaaki. 2010. "Shinkansen gōi no kajitsu: Monju saikai e" [Bullet Train as the Fruit of an Agreement: Toward Restarting *Monju*]. *Asahi shinbun*, April 27.

Reconstruction Agency. 2012. "Zenkoku no hinansha no kazu" [The number of Evacuees all over Japan]. http://www.reconstruction.go.jp/ (accessed 1 September 2011).

Sasaki, Chikara. 1996. *Kagakuron Nyūmon* [Introduction to the Debate on Science]. Tokyo: Iwanami.

Sato, Masanori. 2010. "Yutakana Setonaikai to teiseiseibutsu to genshiryoku hatsuden" [The wealth of Seto Island Sea, Benthos and Nuclear Power Plants], Paper presented at a symposium entitled: "Kaminoseki: the last place where the wealth of the Seto Island Sea remains," Nanzan University, Nagoya, Japan, September 23.

Yakushiji, Taizo. 2011. "What Should We Learn from the Nuclear Crisis?" http://www.jiia.or.jp/en_commentary/201109/28-1.html. (accessed 10 December 2011).

Yasumaru, Yoshio. 2001. *Kindai Tennō-zō no keisei* [Making the Image of the Modern Emperor].

Yomiuri online. 2011. "Shizuoka Motoyama cha no ichibu de kijunchō hōshasei seshiumu" [Tea contaminated with Cesium exceeding the safety limit produced at Shizuoka Moyoyama]. 9 June 2011. http://www.yomiuri.co.jp/national/news/2011 (accessed 30 March 2012).

——. 2012a. "Kikan konnan 25000 nin hinan no sanwari" [25000 evacuees, 30 per cent of all evacuees, have great difficulty in going home]. *Yomiuri online* 7 January 2012. http://www.yomiuri.co.jp/feature/20110316-866921/news/2012 (accessed 26 September 2012).

——. 2012b. "Kaminoseki genpatsu chakkō mitomezu … Chūgokuden wa menkyo shikkō e" [Commencement of Kaminoseki Nuclear Power Plant not accredited … the CEPC will lose its license]. *Yomiuri online* 5 October 2012. http://www.yomiuri.co.jp/atmoney/news/2012 (accessed 13 October 2012).

Yoshioka, Hiroshi. 1987. *Kagaku kakumei no seijigaku* [Politics of the Scientific Revolution]. Tokyo: Chūkō shinsho.

——. 1999. *Genshi ryoku no shakaishi* [Social History of Nuclear Power]. Tokyo: Asahi Sensho.

Yoshizaki, Yoshio. 2001. "Kaminoseki machizukuri renraku kyōgikai" [Kaminoseki Town Management Council]. In *Kokusaku no Yukue*, ed. Asahi shinbun Yamaguchi Branch, 286–288. Kagoshima: Nanpō Shinsha.

FORGIVENESS AS THE OPENING OF THE FUTURE*

Karmen MacKendrick

Abstract: Religiously based arguments in favor of forgiveness are common and are usually based on human imitation of divine models. This article takes a slightly different approach by starting with creation stories to argue that the very structure of time, with its open and unpredictable future, can be read as a form of divine forgiveness that may be accepted or refused. Imitation of the divine model is then the sustaining of openness and promise, based on the re-telling of the past so that it does not foreclose future possibility. This enables us to analyze the conditions that make forgiveness more or less likely and more or less helpful, as well as some of the ways in which forgiveness must interact with justice in the work of peace building and reconciliation.

Keywords: Forgiveness, reconciliation, creation myths, trauma, peace building.

The time of forgiveness is deeply curious, tying origin to the openness of the future. Jacques Derrida asks whether forgiveness is "a power of man—or. . . reserved for God. . ." (*To Forgive* 46),[1] but I suspect we do better to trouble that dichotomy, and I shall do so here by echoing Elliot Wolfson's argument that divine forgiveness is fore-given, given before or already at any time in which we might have need of it (*Fore/giveness on the Way*). It is given *as* a dimension of time, as the future, time's dimension of possibility and novelty—a possibility we too may give as well as receive.[2]

---

* This paper is adapted from a presentation for the ISST conference on Origins and Futures, which in turn drew from a chapter of my book *Fragmentation and Memory*. I am deeply grateful for all the insightful responses and questions that helped me in revising the paper to its present form, and offer particular (not exclusive!) thanks to Gus Koehler and Masae Yuasa. I am also much indebted to very kind and helpful comments from the anonymous reviewers of this article, to Claudia Clausius for her thoughtful and extensive editorial help, and to Elliot Wolfson, who in addition to his own invaluable work and suggestions has alerted me to other helpful material on forgiveness, particularly from John D. Caputo, Edith Wyschogrod, Mary-Jane Rubenstein, and Jacques Derrida.

[1] There are, of course, many thoughtful considerations of forgiveness that ask this question from religious perspectives. Among such works in recent thought, Miroslav Volf's *Free of Charge* is notable for its significance and for its resonance with the present project; though it is more narrowly Christian and Pauline than this paper, he too is engaged in consideration of a kind of divine generosity inherent in forgiving.

[2] For another take on the temporal aspect of forgiveness as "the horizon common to memory, history, and forgetting," see the Epilogue in Ricoeur, 457.

This sense of the *possible* is essential to the understanding of forgiveness, perhaps even more central than the sense of *im*possibility that Derrida so famously attributes to it (*To Forgive; On Cosmopolitanism and Forgiveness*). I want here to explore the implications of this view of forgiven time both as it may allow us a new perspective on human forgiveness and as that perspective may feed into our understanding of the complex processes of political reconciliation. From the notion that in the origin of creation itself time is fore-given, we find that futurity is given precisely *as* what cannot be given-in-advance.

We find forgiveness at work in stories about creation, with their play between beginnings, especially beginnings-anew, and open endings, in which newness can happen. Thinking seriously about this temporal play may have implications for how and why we forgive, and even for when we ought to, if we can say such a thing (forgiveness, I suspect, is particularly and peculiarly unprescribable). Our stories of origin and creation are often stories of return as well. For instance, a pseudo-epigraphic Hebrew variant on the Genesis myth vividly foregrounds forgiveness. In this version, God has repeatedly been led by justice to destroy the world, until finally he generates a sustainable creation:

> [Our world] would have had no permanence, if God had executed his original plan of ruling it according to the principle of strict justice. It was only when he saw that justice by itself would undermine the world that he associated mercy with justice, and made them to rule jointly.

But the need for merciful forgiveness appears prior even to this need to sustain the world. In the same account, we read:

> ... repentance had been created long before, and sinners would have the opportunity of mending their ways .... the Temple service would be invested with atoning power.... Finally, the messiah was appointed to bring salvation, which would put an end to all sinfulness. (Ginzberg 6)

And again: "The grace and lovingkindness of God revealed themselves particularly in His ... saying, 'I shall take man from the place of atonement, that he may endure'" (27). Forgiveness is already possible, but endurance comes out of seeking forgiveness as well as out of the mercy that grants it. Likewise, we know that forgiveness between people is most likely to work where it is a mutual process, and to endure where that process can be sustained. To continue in the same account: "In the beginning ... seven things were created ... [and the last of these was] a Voice that cries aloud, 'Return, you children of men'" (6). Before we even *are*, and so before we are turned away, we are called to return. We are remembered, the promise of remembrance holding open the space to hear the voice calling for return. But when we return, when we in turn remember, it is not only the past we find opened. Forgiveness opens the future too.

Forgiveness is not a direct concern in the more familiar opening lines of Genesis. But the influential interpretation of those lines in Augustine's *Confessions* does, despite the author's own deep anxiety, reveal a sense of an open future possibility in the very formlessness prior to creation. We cannot quite envision the formless; it exists as the slip from one form to another—changeability itself. "Everything mutable implies for us the notion of a kind of formlessness," Augustine writes. Matter goes from form to formlessness to form, changing through formlessness again. Our efforts to picture the formless conjure up only monstrous forms (12.16.28). Yet pure change and pure unchangingness alike come tantalizingly near to God's eternity; first of all things is no thing, nothing formed, but all things' possibility. That the world may be otherwise is its very condition.

The Christian myth of forgiveness, picking up on the idea that "the messiah was appointed to bring salvation," seems to be more temporally linear than these tales of first making. Having effectively condemned itself in its originally sinful progenitor, humanity needs some way to negotiate between unavoidable sin and undesirable damnation. That way is the grand sacrifice of the messiah as an incarnate god. But with the "end of sinfulness" comes not a stasis of bliss, but the newly uncondemned opening of time, the future possible. If this is not simply the deterministic playing-out of a prophecy, then it is a folding-up of time, the future into the fore-given past, the present into the eternal, the promise into the open. So in at least some central Abrahamic myths, forgiveness is truly fore-given. But to forgive or to be forgiven is not to go back to a past or to retreat into an unchanging memory. It is, rather, to transform memory, to open the future, and to perceive as a (merciful) gift the temporal structure of the world.

Forgiveness unbinds time from a precondemned stasis of same-again.[3] It promises the future in all its uncertainty. Unlike the law of justice, the function of which is precisely determination, forgiveness does *not* determine. This can easily make us nervous. Forgiveness emphasizes openness and possibility as attributes both human and divine, while justice emphasizes what must be set in advance; an *ad hoc* law is no law at all. The

[3] Interestingly, the Greek *aphiemi* and Latin *dimittere*, these languages' roots of "forgive," both imply un-binding.

importance of this distinction—and the ways in which forgiveness and justice can nonetheless work together—will become especially evident in this paper's final section.

The Possible and Its Probabilities

I want to hold two propositions in tension (though not quite in contradiction) with one another. First: we often feel that some acts and some people are more forgivable than others, and I think that this feeling is grounded in the realities of forgiveness, circumstances that make it more or less *likely*, and perhaps more or less *useful*. At the same time: forgiveness cannot be earned. If we were so innocent as to merit forgiveness, we would have no need of it. A look at these propositions will help us to understand what use we might make of the origin myths, with their emphasis on forgiveness as that which opens the future and makes endurance possible by making possibility itself. The stories tell us that oddly enough, we *endure* because we can *change*. And it is relevant to forgiveness that the future does change and not simply repeat the past: we can forgive repeated offenses, but for most of us, each repetition makes forgiveness a little less probable.

The first factor in improving the likelihood of forgiveness is memory. Despite the popularity of the phrase "forgive and forget," as Wolfson points out, a matter forgotten neither needs forgiveness nor is available to be forgiven (*Fore/giveness on the Way* 153). "Forgive and forget" is based upon a misperception of memory as somehow constant, unaltered except by its loss, and upon the belief that the past thus remembered is also constant, save in the measure of its distance. This seems to me wholly to misperceive the way that the future is freed. Forgiveness is a change in the *meaning* of memory, a transfiguration rather than a dismissal of the past.[4] It is precisely in this that foregiveness is able to transfigure the future too.

To forgive requires that we remember. Of course, we might agreeably reassure someone worried about having done wrong that there is no cause for worry because the wrong was so slight that we've actually forgotten all about it. This does have some of the benefits of forgiveness, in that it gives

[4] The *meaning* of the past may be modified even if, as Derrida argues, the past itself in some sense cannot (*To Forgive* 32).

a sense of freedom to the one who was worried—but within that relief there may be a slightly letdown feeling as well. The pairing "forgive and forget" may be based too upon our experience with those transgressions that we find ourselves unable to forgive, where the best we can do is to refuse the persistence of memory. It is important that forgiveness is not our only option for moving onward, even if it may be among the most effective, particularly for retaining some relation to the one forgiven. Harold Kushner argues that forgiveness declares: "I refuse to give you the power to define me as a victim.... I don't hate you; I reject you" (186). This is indeed a *kind* of future-freeing—it opens up for the "I" a future unburdened by the memories and anxieties associated with the person who has wronged me. But to say, "I reject you" is not to say, "I forgive you." It is, on the contrary, to insist that no possibility can lie before me-and-you. Forgiveness does not simply open or free up the future; it opens or re-establishes a *relational* future of possible interactions between forgiver and forgiven. It admits of degrees: I may open a very limited relation indeed, or I may embrace a transgressor's return wholeheartedly. That element of relationality, as we shall see, becomes centrally important when we consider the political possibilities of reconciliation as opened by forgiveness.

To be forgiven does not require memory, but is certainly aided by it. That is, it is possible to forgive someone who has no memory of having acted wrongly, but it is very difficult (unless the transgression was truly minor, forgettable in every sense). If I feel that a great wrong has been done, and the one who did it was so unaffected as to have no recollection at all of the deed, then I am not very likely to feel forgiving. And one reason for this is that someone to whom such acts are of no concern has little motivation for avoiding them in the future. Repentance, which aids forgiveness, requires remembrance. Indeed, as Edith Wyschogrod points out, our sense that repentance has some efficacy may make us more likely to change: "The claim that wrongs or sins cannot be expiated may... encourage the belief that there is no reason to avoid their repetition" (161). Repentance suggests both regret and a willingness to exert effort, and these make repetition less probable, thus increasing the likelihood of a future in which the forgiven transgression does not recur. It is helpful to the one who would forgive if the wrongdoer feels guilt, or shame, or unhappiness regarding the injury—not (or not only) because we may feel that these affects are themselves a sort of punishment (such that a form of *justice* has been done by the paying of a psychological penalty), but rather because they argue for the possibility of a *transformation.* They give the transgressor a reason to make the future different from the past. This reason is still clearer where repentance leads to atonement, to an actual effort to make up for the wrong.

There are no absolutes here. It is possible, though immensely difficult, to forgive even those who hold themselves blameless. However, I am unlikely, in forgiving an unrepentant person, fully to reopen an interactive space between us, and more likely to alter the meaning of my memories (which are, of course, a part of that space-between) without structuring my life to include this particular other's continued interactive presence. This restructuring may be of many sorts; we may even forgive the dead, transforming our relation to what we have of their presence: that is, to their memory, and its effects upon us. The effect of such forgiveness is limited, but genuine, as it releases the future from a rigid sense of the past, thus "divinely" restoring futurity and openness. What the elements of probability show us is not that forgiveness can or cannot happen under particular circumstances, but that what makes it easier or harder, more or less likely, is also what makes the future between transgressor and victim more or less open, an openness that may include both hope and fearful uncertainty.

Finally, the odds of forgiveness are improved when this openness is mutual, and mutually accepting of uncertainty. For forgiveness to matter, there must be or must have been some possibility that it would not be granted, or if granted, not accepted. To dwell in the possibility of forgiveness, we must be open, responsive, but without ever knowing what will arrive. When we heed the call to return, we go back to the future, not to an easy comfort that we already know. This sense of response-ability has to do with the *possibility* of sharing, the capacity to attend. In accepting or affirming the openness of the future, we do not obey (rules, or even the summoning voice)—we *respond.*

Mutual response must be dialogical, and forgiveness often requires literally that stories be told and heard. Forgiveness can occur in silence, but it is likely to be limited by the extent to which stories are heard (while we may forgive what we feel we cannot discuss, the very absence of discussion forecloses some possibilities). These are not only the stories of those seeking forgiveness. Such stories may be important as acknowledgements of wrongdoing, but they risk sliding into self-justification or self-pity. Thus in a famous instance, in a display of what Hannah Arendt in the subtitle to her pertinent book vividly calls "the banality of evil," Adolf Eichmann insists that he was only ever doing his job (see esp. chapter 8: "Duties of a Law-Abiding Citizen"). Similarly, Antjie Krog in her account of the Truth and Reconciliation Commission hearings insightfully points out, "But there is also the invisible audience—the imagined audience on the horizon somewhere—the narrator's family, colleagues, the new government. And every listener decodes the story in terms of truth. Telling is therefore never neutral, and the selection and ordering try to determine the interpretation" (107). It is immensely difficult to tell—or even to think!—the truths that portray us as having done harm. We may well want or need the stories of the transgressors, but these do not seem to be the most important part of the dialogue that forgiveness requires. Rather, the most urgent narratives are told by those *from* whom forgiveness is sought.[5] Correspondingly, often the most essential task of one who would be forgiven is listening. To understand the importance of this telling, we might usefully explore the temporal mode of trauma, a nearly perfect contrary to the time of forgiveness.

Trauma and the Unremembered

In psychoanalytic theory, a traumatic "experience" never properly enters into time. Rather, exceeding the capacity of consciousness to register and organize, trauma enters straight into the unconscious, into the a-temporality of repetition. In the classic version of trauma, the event that might have been experienced is repressed from conscious awareness entirely; we loosely speak of lesser forms of trauma, in which the event is perceived, but its affective force is too great to register or process. The time of trauma is that of a particular repetition, an amplified moment: an endless present (which was, however, never presented) re-asserts itself into all other presents—and, unless brought into time as memory, asserts itself as the future. Traumatic "memory" might more accurately be described as a strange forgetfulness in which the body itself cannot forget, cannot let go, but must recreate the past. Only through memory can we find our way back into the openness of time. Trauma repeats because its energy, its anxiety and pain, must somehow be "mastered," made manageable, but

5 "Although some research suggests that forgiving does not necessarily require anything from the perpetrator, other research on forgiveness has shown that harm-doers acknowledging their responsibility for causing harm, apologizing for their actions, and asking for forgiveness contribute to victims forgiving perpetrators" (Staub and Pearlman 222). Consider also Hannah Arendt's observation on the proportionality of testimony: "Eichmann was on the stand from June 20 to July 24, or a total of thirty-three and a half sessions. Almost twice as many sessions, sixty-two out of a total of a hundred and twenty-one, were spent on a hundred prosecution witnesses who, country after country, told their tales of horrors" (223).

without memory it eludes that mastery (Freud § 2). As theorists of reconciliation have noted, traumatic damage is not simply individual but may persist for generations as whole peoples struggle to process the incomprehensible. The transformation from symptomatic drama to narrative memory, from re-enacting to telling, depends upon something crucial: a listener. One has to tell the story, to another. Thus time regains its narrative structure.[6]

There is a risk, of course, that sometimes the listener's contributions may be a bit too directly constructive—such is the worry that surrounded the upsurge in "recovered memories." Even if we are rightly cautious, however, we may say two related things: to begin to surpass trauma is to remember it, and memory's dialogical character argues that to remember is to speak it to someone—and to have one's speaking heard, to encounter and to participate in the creation of a receptive space in which time can become open. Trauma still in its traumatic form cannot be spoken, or heard; it can be seen, but almost always the viewer lacks the knowledge to interpret the dramatic presentation.

I do not mean here to identify trauma simplistically with unforgivenness or unforgivingness, but to show an inversion of temporal structure. The release from trauma is, in common with forgiveness, the freeing of the future. What is released is the grip of repetition, its ability to form the future indefinitely in the shape of the absent past. But if no one hears the story the traumatized tell, the strangely a-temporal "memory" may allow the past to retain much of its grip. The dialogical quality is inseparable from the necessity of memory. We forgive most readily when the conversation is *opened*, and the future created with it.

While trauma demands the absence of memory, forgiveness functions only where we do not forget. Those who seek to be forgiven are helped if they can recall what they have done—the odds of releasing the future and making it different from the past are otherwise poor. Those who seek to forgive must know what offense they are forgiving—and may need to say it or to hear it spoken, sharing the memory to know the material for transformation. Remembering does not mean clinging to the sameness of memory, a risk we run particularly when we hold on to the role of victim.

---

[6] See Wolfson: "Time is indicative of a narrative telling, a diachronic plot whose synchronic crux is open-ended, yielding a rhetoric of temporality characterized by a confluence of repetition and change too complex for simplistic binary opposition, a past determined by a future that anticipates the past as a word spoken in the dialogue between the same and other, the question and response" (*Alef, Mem, Tau* 52).

Didier Pollefeyt, considering particularly the possibilities of forgiveness after the Holocaust, warns that the victim risks establishing a permanent role in victimhood "because that identity sometimes opens an almost inexhaustible 'credit line' of sympathy from others." This tempting strategy, as he notes, leaves both perpetrator and victim with "no exit" (59). Forgiveness cannot hold on to the dramatic or even the narrative identity of a single, fixed tale. It depends on realizing that a story may be retold, subtly shifted, passed back and forth. By retelling the story, by shifting emphasis, by including and excluding new elements, dialogue may promote the forgiveness that changes the ability of the past to form the future in its own image; by remembering and then transforming memory, it breaks open the future.

The Future Revealed: Forgiveness as Revelatory Time

A new past emerges in forgiveness; the old is subject not to forgetfulness as a simple loss, but to change. The old story is retold such that other endings are possible. The past is revealed as what it was, yet newly known, made new by the possibility of novelty. To forgive and to be forgiven open the future not by *losing* the past, but by *changing* it, changing its meaning and thus loosening its grip on the future. This transfiguration releases time from either linear inevitability or cyclical repetition. Once more an image drawn from religious thought may be helpful, here that of *revelation*, in which a transformative change of meaning alters our understanding of the past and opens our sense of the future.

Peter Manchester characterizes religious, particularly Christian, revelation as "a radical and retroactive transformation of understanding" (110). Manchester argues for the time of revelation as "a disclosure space, as a field in which horizons can be opened... configuration and transfiguration of events presented, and not merely as a metric space supporting location and order in sequence" (118). Revelation, in this model, is not the sudden disclosure of some new thing at some particular time, but the transfiguration of time as it is, the emergence of old things as quite other than they were (yet as they were already).

Wolfson likewise notes in Kabbala the revelatory element of forgiveness: "Significantly, the return to origin is marked by the uncovering of that which is hidden" (*Fore/giveness on the Way* 163). Both revelation and forgiveness retell "new" stories about the "same" material, and so make that material, that story, not the same at all. In forgiving, I do not deny that someone has acted wrongly towards me, but I transform the meaning of that action so that its harm is no longer constitutive of our relation—or no longer forecloses the possibility of our being in relation at all. In being forgiven, I am granted that same openness, that the past need not create the present and the future in its own image. Divine forgiveness is ever-present as the simple fact that the future opens before us: enduring, we may be otherwise than we are. The voice calling for return uncovers, reveals, this openness.

This openness of the future through the alteration of the meaning of the past—releasing the future from repetition by re-characterizing the past as forgiven, by remembering that transformation is possible—is forgiven time as well, and is fore-given in the very structure of time as opening onto the future. Humanity is sure to sin eventually, but prior even to this is forgiveness. Forgiveness opens the future by its transformative illumination of the past—it *is* a revelation. If forgiveness is transformative, it is precisely because it does *not* forget, but takes up the hurtful, the evil, the seemingly unforgivable and transforms its meaning, thus transforming the future along with the past. It refuses resentment. Forgiveness is entangled in memory, but memory *not* as repetition: rather memory as transfigured and as essential to transfiguration.

Reservations

Though I hope that I have suggested something of the value of forgiveness, there can be no doubt that the demand for forgiveness generates reservations, both rightly and wrongly. It is easiest to deal with those based upon misconceptions. We tend, for instance, to think that if we forgive, we somehow declare that no harm was done. Unsurprisingly, making such a declaration may leave the forgiver feeling resentful, and the declaration itself may well be dishonest, though made in the interest of peace or good manners. This is particularly so if forgiveness simply lifts the burden of guilt, as if it were an acquittal or declaration of innocence. It then risks becoming, as Pollefeyt claims, a "cheap grace," too readily available to be meaningful (56).[7] Such cheap grace is also at work when we declare that

[7] Pollefeyt goes on to suggest that the Catholic church's failure to make a full confession of its own responsibility for anti-Semitism in the 1998 document "We Remember" may rest on the presumption that forgiveness is widely and easily available, and so in fact may actually keep the church from being forgivable (56).

we—or, better, everyone else—should always forgive every transgression, on the argument that God would so forgive. But the stories of origin and implicit futurity argue otherwise: forgiveness is our openness to new possibility in some given relation, and where that possibility is especially improbable, our option to forgive is at best limited. Again, forgiveness is always possible, but it is not always likely; it is not always wise; and it is usually limited and conditional.

Fortunately forgiveness does not, in fact, imply that the forgiven person was never wrong in the first place. It suggests, rather, that that which was or might have been blamed is still blame*worthy*, but that blame, nonetheless, no longer attaches to it; or that blame attaches but that the penalty for that blame no longer holds. In fact, Maurice Blanchot argues that one ought *never* to forgive, because forgiveness also condemns.[8] That is, we only forgive what is really a transgression (to echo Derrida, only what is in some sense unforgivable), and only when we acknowledge it as such.

The combination of blameworthiness and non-penalizing importantly distinguishes forgiveness from the resolution of guilt. Guilt, as Friedrich Nietzsche noted, may well derive from an economic model, "the contractual relationship between *creditor* and *debtor*" (2.4).[9] Justice demands the repayment of debts. To be forgiven, on the contrary, does not erase, but may indeed increase, my sense of indebtedness. (An exception occurs, of course, if I resentfully feel that I did not do wrong in the first place. This reminds us that forgiveness really does imply that there was a transgression.) To be in need of forgiveness is not to be indebted. It is not even to be accused. It is to be in need of return, to want to be back in good graces, to need some possibility not offered by condemnation, but also not offered by justice. I may make restitution, pay fines, serve time, and still not be forgiven; conversely, I may be forgiven without any of those acts. (As I shall note below, however, the extension of forgiveness into reconciliation does make demands that are closer to those of justice.) Forgiveness must always come as a bonus, in or as an act of absolution. Because it follows the laws neither of physics nor of legislation, it can no more be predicted than it can be demanded. In this, it resembles the idea

[8] See Blanchot: "Do not forgive. Forgiveness accuses before it forgives. By accusing, by stating the injury, it makes the wrong irredeemable" (*Writing of the Disaster* 53). I believe this sense of the irredeemable is similar in Derrida (see *To Forgive* 31).

[9] Nietzsche argues that the creditor's right to extract satisfaction by inflicting pain is in turn influential upon the formation of memory, pain being the best mnemonic, and that our entire notion of future obligation and promise-keeping is grounded upon this contractual relationship (2.4–6).

of grace, which is by definition unmerited (it is gratuitous). But there is no cheapness to it—though, being human, we are likely to continue to try to get it as cheaply as we can.

This worry that forgiveness denies wrongdoing is linked to the further confusion between forgiveness and justice. There is no denying that the establishment of justice is, like remembering and dialogue, strongly *conducive* to forgiveness—and, as I shall note below, it is essential to allowing forgiveness to work in reconciliation. A victim of sustained injustice has no reason to believe that the future is truly open to different possibilities. Pollefeyt argues that "[f]orgiveness presupposes justice" (57), but while I scarcely find this notion nonsensical, I would not extend it to the measure of requirement or presupposition. It is true that we forgive both more easily and more effectively when we aren't facing injustice. The resentment or anger we feel at injustice, and the tendency of injustices to perpetuate themselves, make forgiveness difficult—but "difficult" is not impossible. Nor does justice always make forgiveness easy. We are often surprised by how little "closure" justice brings, but perhaps closure is not all that we need: perhaps we need openings too. This tension is one of the most difficult aspects of forgiveness.

Unlike justice, which must be applied impersonally, forgiveness is personal and unpredictable.[10] What disappears in forgiveness is not, as Wyschogrod notes, possibility (*pace* Derrida), but certainty (161). We cannot demand forgiveness, at least not successfully—not with the same reasonable expectations we can and should bring to a demand for justice. Sometimes we think it has been earned, but we cannot rightly insist that someone must forgive—while we *can* insist, even on behalf of others, that justice be done. We cannot even successfully demand that we ourselves must forgive. The most we can do is to push ourselves to give the world and worldly time its due, to take the future as the gift of possibility and to try to share that gift. At times, it is even the case that we do ourselves and others the most good by *not* forgiving: we easily envision, for instance, the disastrously hopeful return of the abused to the persistent abuser.[11]

---

[10] See Derrida: "In a certain way, it seems to us that forgiveness can only be asked or granted 'one to one,' face to face, so to speak, between the one who has committed the irreparable or irreversible wrong and he or she who has suffered it and who is alone in being able to hear the request for forgiveness, to grant or refuse it" (*To Forgive* 25).

[11] Mary-Jane Rubenstein considers such a possibility—that of requiring that one forgive an abuser—as potentially implied by Derrida's odd notion of perfect forgiveness (83–84).

Some (More Hopeful) Implications for Piety, Politics, and Persons

As I have noted, appeals to religious grounds are common in arguments favoring forgiveness. Christianity especially (but not uniquely), with its somewhat unusual version of a divinity made human, tends to view ideal human behavior as modeled upon divine behavior. This would seem to suggest that humans too should hold the future absolutely open, should always forgive every transgression. Indeed, this modeling is a very common argument for the value of forgiveness, and such forgiveness is always possible—but again, it is not always either likely or wise. To insist upon it regardless of circumstance seems to assume that one party's forgiveness assures that the other party too will seek an open and different future. One person can certainly transform a relation, but this will not necessarily bring the other party's actions into line with the forgiver's hopes or needs for transformation. In this last section, I look at reconciliation, in which the need for mutuality is especially clear—and particularly urgent.

I understand reconciliation here in its political sense, as the work of creating a mutually sustainable future between groups with a hostile and often traumatic history between them. Reconciliation is visible; it is the change in social and political structures after inequities on the vast scale of war or apartheid, when that change is at least meant to work to the benefit of both victims and transgressors. Thus it is closely connected to the opening work done by forgiveness, but it must go beyond the personal scale, and it must actualize some of the very possibilities that forgiveness opens. The distinction between reconciliation and forgiveness is both subtle and significant. How that distinction is understood depends especially on how forgiveness is understood. Let me offer a few points that seem especially relevant. First, the effectiveness of forgiveness need not be external to the forgiver: certainly *more* possibilities are opened and held open when the forgiven respond with attitudinal or behavioral changes, but the forgiver's release from repetition and fixity into futurity is the element constitutive of forgiving. Reconciliation, however, must be effective in the world between groups of persons in a way that can be recognized even by an external observer, as well as by both parties. Thus when Viken Yacoubian asks what reconciliation and forgiveness might look like in the context of the Armenian experience of genocide, he notes, "One distinction that has been proposed posits that reconciliation implies restoration of a broken relationship, thereby requiring goodwill from both parties.... Forgiveness, on the other hand, can occur unconditionally, without reconciliation necessarily following it" (224). This mutuality means that reconciliation must be literally dialogical, while forgiveness does not have to be. But it is helped by dialogue, as Barbara Tint points out in her argument for the centrality of dialogue to reconciliation: "forgiveness work may not necessitate dialogue, but reconciliation will almost always require some space in which parties come together to heal the past and imagine a different future.... Reconciliation is a process, like dialogue, which focuses on relationship as the transformative dimension in peace building" (274). This dialogue need not imply friendliness; willingness to participate in the process is what counts.

On a related note, reconciliation must *sustain* the openness of the future and must not return to the patterns of the past. Again, this is something that forgiveness hopes to do, but as I have briefly noted, a transgressor may repeat the transgression without causing the forgiver to retract the forgiveness. Reconciliation, however, has *failed* if it is not sustained in its mutuality.

Reconciliation works at creating and sustaining justice through the transformations offered by forgiveness. It entails not only forgiveness but also a response that sustains the possibility of newness: atonement that we may endure.[12] In the most difficult cases, however, we may need to endure without the evidence of atonement that we continue to desire, as I shall note below.

An example of a failed attempt to receive the forgiveness of a wronged group will actually help us to begin the consideration of reconciliation that works. In his famous essay "The Sunflower," Simon Wiesenthal tells the story of a dying Nazi soldier who seeks forgiveness from him for actions harmful, even fatal, to other Jews. This is a request halfway between individual forgiveness (the soldier asks only for himself) and group reconcili-

---

[12] Here too there is something of a religious parallel: in some versions of the Christian tradition, that mutuality is what characterizes human redemption: damnation is simply the stubborn and angry refusal of divine openness. The major difference, of course, is that such redemption does not require justice for its sustaining. See Gilles Deleuze on Leibniz's unusual theory of damnation: "... the damned, Judas or Beelzebub, does not pay retribution for a past action but for the hate of God that constitutes the present amplitude of his soul and fills it in the present. He is not damned for a past action, but by a present action that he renews at every moment" (71). Cf. *Julian of Norwich*, as Denys Turner argues: "For Julian... what is 'real' is the divine love. Sin, in being the refusal of that love, is the refusal of reality... As Julian says, peace is 'always in us.' And the only thing that can separate us from that peace is the 'wrath' that is only 'on man's side,' not at all on God's" (93). In this, both Julian and Leibniz are odd descendents of Augustine, for whom evil is privation or nothingness; to accept privation is simply to refuse the ever-present divine love, and the punishment for sin is not some after-effect imposed by a distant divinity, but inherent in sin itself as the loss of joy. See *Confessions*, passim throughout.

ation (yet he asks from or on behalf of all Jewish people). Wiesenthal does not offer forgiveness, and perhaps cannot, but the question of whether he *should* not has been fiercely debated. At issue is not only whether Wiesenthal should himself be forgiving, but also whether forgiveness may be offered on behalf of others, or in the name of others (we might also ask whether forgiveness may be *sought* in the name of others, such as one's wrong-doing ancestors or unethical members of a group with whom one is identified). In the fullest sense, I cannot forgive on behalf of someone else because forgiveness must consider persons;[13] it opens up a very particular part of the future, a relation, and so just who is related is crucial. But I legitimately forgive harm done to others insofar as that harm has an impact on me, whether that impact is immediate (such as that done to a loved one) or more distant (as in a lingering unjust social structure). That is, I can forgive wrong I suffer because others were harmed—most obviously others who are my cohort or forebears. This does not mean that I should, still less that I must, forgive in these instances, but only that such forgiveness is possible. I may forgive the actual wrongdoers, altering their place in my memory; I may forgive their descendents if they seek it, acknowledging less traumatized possibilities between us—and opening the possibility of a more just future. It may very well be that law only works, as a practical matter, because it is sometimes tempered by forgiveness; it may be that pure justice, as Derrida claims, is impossible and only imperfectly instantiated in law ("Force of Law"). But this complex relation does not function to bring forgiveness under law, nor, of course, vice-versa. Of course, justice may obtain between persons, and it may be a matter of ethical and not simply of juridical law, but justice respects all persons without differentiation; it cannot afford to be too responsive.

Forgiveness is personal in being responsive to the individuality of both forgiven and forgiver (thus I forgive only for me). It may also be that Wiesenthal cannot forgive because the injustice before him is still too great—which must sound odd, as I have said that forgiveness may function as mercy even contrary to the demands of justice. This sense of the need for justice when we work on the larger scale moves us toward the way in which reconciliation, involving groups and thus always working on behalf of more than one self, demands justice as well.

Such difficulties for justice and forgiveness alike occur especially with large-scale and long-term wrongdoing that may go back for generations.

---

13 On law and grace see Blanchot, *The Step Not Beyond* 25–26.

Such wrongs tend to become institutionalized into structures. It is hard to see exactly how forgiveness for a structure might work, or how it might effectively open the future if it were somehow to be granted. These long-term harms call for justice, and we are obligated to seek justice where we can—where, that is, some balance may actually be made or restored by it, some victimization undone, and more open possibilities restored. In an echo of our opening creation story, though, justice alone may no more suffice than forgiveness would by itself.

The upsurge of work on reconciliation in recent decades tends to be sociological or practical. Theorists may offer definitions for working purposes, but where they theorize and analyze, they are primarily concerned with working out how best to reconcile and how to retain the gains of reconciliation. Many work from, and write about, particular historical and political contexts, or about reconciliation across several such situations, with a focus on method. In one such analysis, Antoine Rutayisire writes of the struggle for reform in Rwanda: "During most of 1994 and 1995, everybody thought justice would be our healing solution. By 1996, we already knew it would not work! That is when 'reconciliation' became the word of the day" (183). Coming out of a wish for justice and the hard realization that justice will not be realized without mercy, reconciliation reminds us that there are conditions in which the *interplay* of forgiveness and justice is essential, and that the view of forgiveness as the creation and acceptance of an open future may offer us some interesting ways to interpret this complex process. Distant though we are from creative divinity, we too run the risks of destroying our (political) worlds if we are too strict about the law—and this is precisely because we believe in and want to sustain its rule. Justice alone may be too unyielding, but forgiveness alone is too indeterminate to work on the political scale.

The two may work together, however, and in so doing remind us again that conditions conducive to forgiveness may be essential to reconciliation. In his careful consideration of the needs of perpetrators and victims in serious crimes (his exemplar is murder), Raymond Paloutzian points out one important ground for this mutuality:

> [A] victim is more likely to forgive to a greater degree if the offender admits to having committed the crime [and] . . . if he or she feels safe from future harm. This is more likely if there are public safety policies and procedures in place to ensure that the victims are safe. . . . [B]ecause a crime is both a violation of a public law and an act against individuals, forgiveness is easier if the victims see their government initiate a process of restitution and reconciliation. (79)

As this need for restitution indicates, reconciliation with its inclusion of justice is not simply forgiveness on a larger scale. Increasingly, however, theorists of reconciliation emphasize the very openness of the future that forgiveness makes possible. Influential theorist John Paul Lederach makes futurity central to his conception of reconciliation: "Reconciliation, in essence, represents ... the point of encounter where concerns about both the past and the future can meet" (*Building Peace* 7; see also 112). Like forgiveness, reconciliation requires remembrance while refusing to define the future by the past. "Reconciliation usually requires coming to terms with the past, but doing so in a manner that will promote a new political culture and commitment to a shared future" (Chapman, 258).

Like forgiveness, reconciliation demands the very unpredictability that may make futurity frightening. In reconciliation, the unpredictable element is especially that of human relations. Shifting the focus from particular desired *outcomes* and a sense of resolution to the sustaining of relational *process* has been central to recent work on reconciliation (see especially *Building Peace* 24, 150). Lederach argues that we must "seek a genuine and committed relationship rather than results" ("Five Qualities" 203), in a process that begins in "a reorientation toward the centrality of relationships" (195). The need for mutuality is emphasized if the process makes the satisfaction of each party dependent upon that of the other, in a way that sustains the openness of that relation: "[R]econciliation must envision the future in a way that enhances interdependence." Such an envisioned future requires a break from the cycles of the past, and this requires genuine dialogue: "Acknowledgment through hearing one another's stories validates experience and feelings and represents the first step toward restoration of the person and the relationship" (*Building Peace* 26–27). Here again, reconciliation *requires* what *helps* forgiveness. That is, reconciliation cannot proceed without dialogue and acknowledgement. Forgiveness can—but without them, it is less likely.

Like forgiveness, reconciliation is opposed by traumatic persistence—and difficult though it is to work through a personal trauma, the difficulty of working through a mass trauma is far greater. Rutayisire points out some of the ground for this, even when we know that such release might be valuable: "Trauma reduction and healing play a great role in the process of forgiveness and reconciliation," he writes. "Wounds that have been inflicted on us, particularly through mass tragedies like a genocide, tend to be 'sacralized' and people like to hold on to them, not daring to desecrate them. This refusal to let go ... is a sure road to an eternal state of victimhood" (185). Again, the future is more truly open where we have both reason to believe that it will not repeat the past and the ability to release the future from that past's determination. A person for whom victimhood becomes essential to identity cannot open a possible future with the victimizer without risking his or her very sense of self—a self defined by a damaging trauma. Something parallel may occur at the larger scale: whole peoples may come to define themselves, even if this is only one aspect of their self-definition, by these extraordinary destructive events, likewise resisting the opening of possibilities. Such traumas become "sacralized," and where we hold something to be sacred, we are particularly unwilling to consider other possibilities. Reconciliation demands that those possibilities be actualized. We may forgive and hold open the future and turn out to be mistaken in our hopes that the future will not repeat the past. But if the future does repeat the past, reconciliation, unlike forgiveness, has failed.

It is here that forgiveness in its very otherness from justice is important to reconciliation. We can easily fall into a persistent victim status when we insist on justice for what cannot be balanced or closed. We must be wary of thinking that we can open the future by closing the past. Where time, death, sheer scale, or any other factor makes justice unlikely, we do ourselves no favors by insisting that the future remain foreclosed by trauma. And that foreclosure happens too easily when we think that justice will bring some sort of "closure," some sense of completion and distance, to the past, or will balance out some transgression that cannot after all be set right.

We must be careful, that is, not to let our desire for closure and completion, which will be all the stronger where circumstances are ugliest, turn into our foreclosure of the future. If we have been harmed, we must be able to acknowledge this without sacralizing our victimization. Likewise, if we as victimizers are granted amnesty (as may happen when justice alone proves unworkable), we must be careful not to let reconciliation lead us into the smug assumption of our own innocence. Amnesty is a form of political or judicial forgiveness, and the crimes it does not punish as law might allow may be immense. But it is not personal forgiveness, nor is it affective (though one hopes it is effective). Like forgiveness on a personal level, amnesty still holds a crime to be blameworthy, to merit the punishment it will not receive. Amnesty is, at least ideally, granted only when the destructive power of justice for the past would be obliterative of the possibility of a just future; when the very scale of the crime, which seems to demand intense and lasting punishment, means that punishment would take one side (a people, a party) out of any possible reconciliatory process. Not all processes of reconciliation involve this tricky step, of course. When they do, as perhaps they sometimes must, the process becomes more difficult still, demanding as it does openness and dialogue with those one may nonetheless regard with lasting personal loathing. This is not so distant from forgiveness as it might seem. An individual is unlikely to loathe those whom she or he has forgiven, but may very well not *like* them, even though the past is no longer relationally formative. The possible dislike or hate between political or social or ethnic enemies is likely to be more lasting and more profound, but in this case too, the affect is neither wholly irrelevant (it demands greater care in dialogue, for instance) nor determinative of outcome. We need not like one another to work together; but we do need to agree on envisioning a workable future. Once more, that sustained open future is what most matters.

Reconciliation will only work with the effort of both parties, with the *actual* alteration of the future from the patterns of the past, and with the inclusion of some measure of justice alongside forgiveness, even in cases in which just punishment is not imposed for all past crimes. Reconciliation must combine forgiveness and justice, even while acknowledging the difficult tensions between them. It is justice, with its attention to fairness and equity, that allows *sustaining* of the opening that forgiveness, whether individual or communal, can *create*. Reconciliation reminds us that to make the possibilities of forgiveness more than individual and more than psychological, work must be done and redone—work that is both grounded in forgiveness as the opening of the future and dedicated to sustaining that futurity.

In forgiveness the fragments of memory mend together into a new narrative to break open the future, mend and break again, constantly retelling and making retelling possible. It may be effective for one who forgives, but will be more so, and more widely so, if that openness is also sought by those forgiven. In the political or social realm of reconciliation, that mutuality becomes a requirement. Forgiveness is the revelation of the future in its openness instead of the traumatic, damning closure of repetition—whether that future is my own or one shared with an entire country. It reminds us that the world itself can come to be only as and through the possibility of being otherwise, can have its future only when we remember its origin in the openness of time.

References Cited

Arendt, Hannah. *Eichmann in Jerusalem: A Report on the Banality of Evil.* New York: Penguin Classics, 2006.

Augustine. *City of God.* Trans. Henry Bettenson. New York: Penguin Classics, 2003.

——. *Confessions.* Trans. Henry Chadwick. Oxford: Oxford University Press, 1991.

Blanchot, Maurice. *The Step Not Beyond.* Trans. Lycette Nelson. Albany: State University of New York Press, 1992.

——. *The Writing of the Disaster.* Trans. Ann Smock. Lincoln: University of Nebraska Press, 1995.

Chapman, Audrey R. "Truth Commissions as Instruments of Forgiveness and Reconciliation." *Forgiveness and Reconciliation: Religion, Public Policy, and Conflict Transformation.* Ed. Raymond G. Helmick, SJ and Rodney L. Petersen. Philadelphia: Templeton Foundation Press, 2001. 257–277.

Deleuze, Gilles. *The Fold: Leibniz and the Baroque.* Trans. Tom Conley. Minneapolis: University of Minnesota Press, 1993.

Derrida, Jacques. "Force of Law." Trans. Mary Quaintance. *Deconstruction and the Possibility of Justice.* Ed. Drucilla Cornell, Michael Rosenfeld, and David Gray Carlson. New York: Routledge, 1992. 3–67.

——. "To Forgive: The Unforgivable and the Imprescriptible." *Questioning God.* Ed. John D. Caputo, Mark Dooley, and Michael J. Scanlon. Bloomington: Indiana University Press, 2001. 21–51.

——. *On Cosmopolitanism and Forgiveness.* Trans. Mark Dooley and Michael Hughes. London: Routledge, 2001.

Freud, Sigmund. *Beyond the Pleasure Principle.* Trans. James Strachey. New York: W.W. Norton & Co, 1961.

Ginzberg, Louis, ed. *The Legends of the Jews.* Vol. 1. Web. 12. Sept. 2012. <http://philologos.org/__eb-lotj/vol1/one.htm#1>.

Krog, Antjie. *Country of My Skull: Guilt, Sorrow and the Limits of Forgiveness in the New South Africa.* New York: Three Rivers Press, 2000.

Kushner, Harold S. Response. *The Sunflower: On the Possibilities and Limits of Forgiveness.* Text by Simon Wiesenthal with symposium of responses and discussion. Ed. H.J. Cargas and B.V. Fetterman. New York: Schocken Books, 1998. 183–185.

Lederach, John Paul. *Building Peace: Sustainable Reconciliation in Divided Societies.* Washington, DC: U.S. Institute of Peace Press, 1997.

——. "Five Qualities of Practice in Support of Reconciliation Processes." *Forgiveness and Reconciliation: Religion, Public Policy, and Conflict Transformation.* Ed. Raymond G. Helmick, SJ and Rodney L. Petersen. Philadelphia: Templeton Foundation Press, 2001. 193–203.

MacKendrick, Karmen. *Fragmentation and Memory: Meditations on Christian Doctrine.* New York: Fordham University Press, 2008.

Manchester, Peter. "Time in Christianity." *Religion and Time.* Ed. A.N. Balslev and J.N. Mohanty. Leiden: Brill, 1993. 109–137.

Nietzsche, Friedrich. *On the Genealogy of Morality.* Trans. Maudemarie Clark and Alan J. Swensen. Indianapolis: Hackett Publishing, 1998.

Paloutzian, Raymond F. "The Bullet and Its Meaning: Forgiveness, Nonforgiveness, and their Confrontation." *Forgiveness and Reconciliation: Psychological Pathways to Conflict Transformation and Peace Building.* Ed. Ani Kalayjian and Raymond F. Paloutzian. Dordrecht: Springer, 2009. 71–80.

Pollefeyt, Didier. "Forgiveness after the Holocaust." *After-Words: Post-Holocaust Struggles with Forgiveness, Reconciliation, Justice.* Ed. David Patterson and John K. Roth. Seattle: University of Washington Press, 2004. 55–59.

Ricoeur, Paul. *Memory, History, Forgetting.* Trans. Kathleen Blamey and David Pellauer. Chicago: University of Chicago Press, 2004.

Rutayisire, Antoine. "Rwanda: Repentance and Forgiveness—Pillars of Genuine Reconciliation." *Forgiveness and Reconciliation: Psychological Pathways to Conflict Transformation and Peace Building*. Ed. Ani Kalayjian and Raymond F. Paloutzian. Dordrecht: Springer, 2009. 181–187.

Rubenstein, Mary-Jane. "Of Ghosts and Angels: Derrida, Kushner, and the Impossibility of Forgiveness." *Journal for Cultural and Religious Theory* 9.1 (2008): 79–95.

Staub, Ervin and Laurie Anne Pearlman. "Healing, Reconciliation and Forgiving after Genocide and Other Collective Violence." *Forgiveness and Reconciliation: Religion, Public Policy, and Conflict Transformation*. Ed. Raymond G. Helmick, SJ and Rodney L. Petersen. Philadelphia: Templeton Foundation Press, 2001. 205–227.

Tint, Barbara S. "Dialogue, Forgiveness, and Reconciliation." *Forgiveness and Reconciliation: Psychological Pathways to Conflict Transformation and Peace Building*. Ed. Ani Kalayjian and Raymond F. Paloutzian. Dordrecht: Springer, 2009. 269–285.

Turner, Denys. *Julian of Norwich: Theologian*. New Haven, Conn.: Yale University Press, 2011.

Volf, Miroslav. *Free of Charge: Giving and Forgiving in a Culture Stripped of Grace*. Grand Rapids, Mich: Zondervan, 2006.

Wiesenthal, Simon. *The Sunflower: On the Possibilities and Limits of Forgiveness*. With symposium of responses and discussion. Ed. H J. Cargas and B.V. Fetterman. New York: Schocken Books, 1998.

Wolfson, Elliot R. "Fore/giveness on the Way: Nesting in the Womb of Response." *Graven Images* 4 (1998): 153–169.

——. *Alef, Mem, Tau: Kabbalistic Musings on Time, Truth and Death*. Berkeley: University of California Press, 2006.

Wyschogrod, Edith. "Repentance and Forgiveness: The Undoing of Time." *International Journal for Philosophy of Religion* 60 (2006): 157–168.

Yacoubian, Viken. "Forgiveness in the Context of the Armenian Experience." *Forgiveness and Reconciliation: Psychological Pathways to Conflict Transformation and Peace Building*. Ed. Ani Kalayjian and Raymond F. Paloutzian. Dordrecht: Springer, 2009. 223–235.

INDEX

Abraham 181–182, 261
acedia 228–229
action 175–184, 186–187
  social action 8, 55
activism 199
afterlife 116
Against the Day 71, 73
agape 190, 196–197, 201, 204
aggression 190, 198, 202
  aggression through the institution of slavery 195, 196
Allegory of the Cave 173
Alvarez, Walter 28, 139, 150
Alvarez hypothesis 150
amnesty 276
An Inspector Calls 118, 120
anachrony 85
anxiety 107, 175, 209, 250, 261, 265
apartheid 271
apocalypse
  Apocalypse 184, 244
  apocalyptic literature 172, 184–186
  apocalyptic violence 172
  apocalyptic utopian visions 172
Apollo space program 126
applied science 124
archē 237, 244, 249–250, 254, 256n.
Archosaur 130
Arendt, Hannah 170, 172, 175–182, 186, 217n., 264–265
Aristotle 93, 98, 183, 213–215
Arthuriad 58
artistic beauty 196
asymmetry
  past-future asymmetry 139–140, 142, 148–149, 151, 153–156, 159–164
atom
  atomic energy, see nuclear energy
  letters as atoms 34
  periodic table of atoms 34
Atoms for Peace 247
atonement 260, 264, 272
  Atonement 71
Augustine 31, 140, 261, 272
Auschwitz 179
Austen, Jane 84
Auster, Paul 71–72
autonomous selfhood 190
avant-garde 199, 222, 232

Bachelard, Gaston 190–191, 196–198, 204
  Bachelard on Bergson 191
Bakhtin, Mikhail 53
basic science 7, 123–124, 126, 130, 135
Baudelaire, Charles 213–214, 218, 222, 225, 231
Beckett, Samuel 105, 214
Beethoven, Ludwig van 20
beginning 1, 9, 11, 25, 29, 31–32, 34, 83–85, 111, 115, 230
  beginning and / or end(ings) 38–39, 41, 97, 142, 224, 260–261
  beginning of the universe 152, 155n.
  beginning presupposing end 88–92
  consecration of beginnings 198
  death is return to beginning 107–108
  principle of beginning 175
  original beginning (see also *archē*) 244
  origin before beginning 220
  renewal of beginning 45
  reversing endpoint and beginning 81
behaviorism 177
being
  pure being 32
  Being (Heidegger) 37, 47
Being and Time 37
Benjamin, Walter 98, 187, 215–228, 232
  Benjamin and Kraus 215
Beowulf 62–63, 73
Bergson, Henri 53, 70, 189–190, 197–199, 201–212, 204
Bergsonism
  discontinuous Bergsonism 189, 192, 193
bi-location 71
Big Bang 30, 125
Big Science 123–125, 135
  Big Science Projects 124
  applied science
    ITeR Fusion Reactor 126
    Space Shuttle 126
  basic science
    exobiology 125
    Human Epigenome Project 123, 125

Human Genome Project 123, 125, 127–130, 132–133
Webb Telescope 126
biography 42
biology 176
basic biology 128
developmental biology 127
biomedical sciences 128
biotechnology 128
birth 51, 53, 61, 175–177, 181
birth of the universe 173
black lives 191
Blanchot, Maurice 269
blind faith 195
Bloch, Ernst 93
Block Universe 140–145, 147–149, 152
Blumenberg, Hans 38
Boas, Franz 192
body 41
boundary
boundary of the universe 51
brain
human brain 173
branch-dependence 154, 159–160
Brecht, Bertolt 98
Buddhism 201
Burns, Robert 36

calendar 105
Callender, Craig 116
cancer 127–128
Canterbury Tales 55
carbohydrate metabolism 133, 135
Caxton, William 63
causality 59, 91
causality jumps in fiction 87
causality of tragic effect 93
causality replaced by temporal sequence 92
mechanistic causality 191
suspension of causality 192, 198
causation 50–51
cause and effect 51
Chang, Belinda 130
chaos 32, 50, 70–71, 91, 174
primeval chaos 32
Chaucer, Geoffrey 55
Chekhov, Anton 94
chivalry 59
chloroplasts 134
chorismós 39
Christianity 271
chronology 52, 62
chronological order / sequence 63, 66–67, 110
dendrochronology 28–29
master chronologies 29
chronometrics 32
chronometric revolution 27
chronotopes 53
Chuang Tzu (= Zhuangzi) 31
Churchill, Winston 16
clinamen 35
Clinton, Bill 128
closure 83, 221, 230, 232, 270, 276–277
narrative and closure 232
coherence (see also decoherence) 95, 159, 207, 224
coherence vs. terror 224, 229, 232
coherent sense 69
commitment to coherence 218
narrative coherence 87
modes of narrative coherence 61–62, 65, 68, 70, 73
Coleridge, Samuel Taylor 24
Collins, Francis 127–128
Communist International 20
community 179
community of equals 180
political communities 179
community and time 243–244
complementarity maintenance 196
complexity 24, 40, 129, 170, 173–175, 184
boundary of complexity 173
temporal complexity 171
conflict 23, 170–175
creative conflict 23
final conflict 20, 23
unresolvable conflicts 24
consciousness 53, 58–67, 69, 72
community consciousness 120
consciousness of temporal existence 38
content and the object of consciousness 46
cultic consciousness 45–47
dream consciousness 111, 116
Halliday's world of consciousness 59
individual consciousness 51, 61
pure consciousness 200
stream of consciousness 69
time consciousness 111
world of consciousness 58–63, 67, 72
constitution (of objects) 39–40
cosmology 29–32, 140
cosmic singularity 32
cosmology as a game of detection 30
mathematical cosmology 30
creation 173–175
Creation 174, 260

creation of man 175
creation of the universe 30, 33
creation myth 29
creation story 34, 260, 274
origin of creation 260
poetic creation 214–215
primordial creative events 35
creativity 90, 145, 174, 195
expressions of creativity 195
love essence of the creative effort 197
Cretaceous-Tertiary K-T extinction 27–28
crowd 225
Man of the Crowd (Poe) 225, 230
motif of the crowd 225
movement of the crowd 225–226
urban crowd 225–226
Cold War 247, 250
Country (Japan) built on Science / Nuclear Power 246, 247, 249, 250, 255

Dada 222–223
Dada collages 222
Dangerous Corner 114
Daoism 31
Dasein 37, 39, 47
Daseinsrelativität 40–41, 45
dateline
dateline as origin of a narrative 26
detecting datelines 26, 35
death 42, 51, 53, 61, 67, 106–107, 169–170, 173, 176, 186
heat death 152n, 154n.
decoherence 154, 162
decoherence (definition) 154
decoherence and branch dependence 154
decoherence and irreversibility 154
decoherence and probability 160
decoherence leading to increase in entropy 160
Defoe, Daniel 64
déjà vu 107, 109
Deleuze, Gilles 53, 272
DEMETER 133, 135
dénouement 94, 98, 224, 232
dénouement as origin (Poe) 88
Derrida, Jacques 259, 269–270, 273
Descartes, René 39n., 57–58
detective fiction 26, 86, 93, 95, 97–98
detective fiction seen by literary theorists 98
productive tension between origin and ending in detective fiction 99
Sayers's characterization of detective fiction 98
dialogue 54n., 63, 265–267, 270, 272, 275, 277
Dickens, Charles 89
diegesis (see also: narrative) 59–60, 82
discourse 46, 72n., 80–81, 86, 95, 209–210, 213
discours 53
disruption 182, 185
divine quaternity 202
divine retribution 195
DNA 134
DNA methylation 131
DNA molecules 127
DNA packaging 130
DNA sequence 127–128, 131
structure of DNA 127
doppelgangers 71
drama(s) 83–86, 93, 109–112, 190, 211, 266
analytical drama 84–85, 93
dreams 103, 106, 110, 112, 117, 202
dreams giving knowledge of the future 109
dreamtime 116
pre-cognitive dreams 111
dualism 40–41, 52, 72n.
Kantian dualism 40, 41
Dunne, J.W. 110–111
durability 177
duration 189–190, 193, 203
behavior initiating duration 190
continuous duration 192
disposing oneself to duration 195
duration as time separate from space 190
experience of duration 196
generation of duration 197–198, 200, 204
mantra to invoke duration 196
plurality of durations 191
pure duration 190
recuperative duration 201
self-generated duration 196
durée, see also duration 53, 189

Eco, Umberto 98
economy, economics 177
political economy 177
economics as rationale for nuclear power 240
electro-magnetic radiation 51, 170
Eliade, Mircea 243
Eliot, George 66
Eliot, T.S. 129, 232
endo-parasites 134
entropy 140, 146, 154

decoherence leading to increase in entropy 160
entropy and order / disorder 156
entropy and past-future-asymmetry 162, 163
entropy and time travel 163
epigenesis
epigenetic imprint 131–132, 134
epigenetic packing 135
epigenome 125, 131–133
Epigenome Consortium 130–131
epistemology 69
eschatology 170, 172, 182, 184–187
e. and apocalyptic violence 172
eschatological movements 172
esoteric thought 106
eternal recurrence / return 107–108
eternity 108, 173–174, 176, 186, 256
eternity as repetition 108
evil
radical evil 178–179
evolution 108, 125, 129–135, 159, 193
evolution of language 59–60
evolutionary schema 176
organic evolution 50, 60, 64
temporal evolution 173–175
theory of evolution 175
evolutionary consequence 135
existence 31–32
collective existence 45
human existence 52, 127, 170, 176–177
personal / individual existence 41–43
temporal existence 38, 46, 172, 175–176, 184
exobiology 123, 125, 133–134
extinction of dinosaurs 150

fabula 53, 80
fantasy 51, 113, 210, 221
fast breeder reactor (see also: nuclear energy) 240–241
fate 83, 94, 104, 107
fear 65, 120, 174
Fear and Trembing 182
feeling(s) 56–57, 61, 63, 195, 253, 262–263, 275
feminine principle 202
fiction 77–101, 105, 142, 189, 208–210, 214–215, 221, 232
detective fiction 26, 93–97
fictional reality 68
origins and fiction 86
truth of fiction 223
field 54–55, 58
field of social action 54
fields of being, doing, sensing 58–59
Fielding, Henry 66
finiteness 169
finitude 42, 175–176
forgiveness 178–179, 259–276
Christian myth of forgiveness 261
divine forgiveness 259, 268
forgiveness also condemns 269
forgiveness and justice / law 270, 273–274, 277
forgiveness and memory 262–263, 277
forgiveness and stories of creation 260
forgiveness opening the future 261, 263
reconciliation and forgiveness 271–272
time of forgiveness 259–260
unpredictability of foregiveness 270
Fraser, J.T. 29–32, 49–54, 59–61, 64, 67–72, 121, 170–176, 181–182, 184, 187–188
free act 190, 199
free choice 189
freedom 24, 170, 173–178, 181–182, 186, 218
freedom to do nothing 200
French Renaissance 28
future 79, 82, 87, 107, 172, 179–181, 184, 186–187
a dreamed past in a dreamed future 211
ability to access the future 112
ambitions of the future 209
foreclosure of the future 276
future as resolution of narrative 79–80
future memory 67
future past tense 80
future-vision, watching the future 111, 192–193, 196, 200
futurity 260, 264, 269, 271, 275, 277
intradiegetic future 83
openness of the future 259, 264, 268, 272, 275
origin defining future 220
revolutionary future (Benjamin) 215
structure implied by origin and future 80
suspense and future 82
uncertainty of the future 82
virtual future 87

Game, The 84
gender 196, 200
female time-experience 198
feminine ways 196
male-female reciprocity 196

genes 129
families of genes 129
gene activity 131
gene regulation
carbohydrate metabolism 133
methylation and imprinting 131
Genesis 34, 181–182, 260–261
Genette, Gérard 52, 59, 85
genocide 183, 275
Armenian experience of genocide 271
genomes 129, 130, 134
evolution of genomes 130
Neanderthal genome 130
parental genomes 132
genre(s) 61, 213–215
accounts of origin as genre 29
detective / mystery genre 86, 93, 95n., 98–99
genre conventions as advance knowledge 83
genre conventions of mystery 94
Geworfenheit 187
Ghandi 200
ghost 18
Ginzburg, Carlo 98
grace 270, 273n.
cheap grace 268
grace and divine kindness 260
Grass, Günter 85n., 89, 90
Great Time, The 105
guilt 86, 119, 245, 263, 268–269

half life 51
Halliday, M.A.K. 49, 53–72
hamartia (hero's tragic flaw) 83
Hamlet 207, 229n.
hand-dials 105
Harlem Renaissance 191
haunting 178
health care 128
Hegel, G.W.F. 172
Heidegger, Martin 37–40, 42–44, 47, 174 187
Heilbrun, Carolyn 98
hermeneutics 42, 44, 98
Hirohito, Emperor 250–251
Hiroshima atomic bomb 235, 242, 244–245
histoire 53, 80
historiography 213, 215, 219–220, 222, 225
historiography and *acedia* 229
historiography as *musisch* 215, 221
history 27, 47, 71, 141, 146, 179, 186–187, 207, 212–224, 228, 232, 242, 244
alternative histories and branch dependence 160–161
Art History Mystery 28
Big History 27
future history 143, 161
history and sociotemporality 61–62, 68
history and institutionalization 242
history of cosmologies 30
history of the universe 174
philosophy of history 172, 186
traumatic history 271
Hitler-Stalin Pact 216
Homer 63, 98, 143
hope 170, 173–174, 180, 182–188, 216, 233
human spirit of hope 228
hope for order 232
hope and time imagery 244
forgiveness and hope 270–272, 276
hors-temps 190, 198
hour glass 105
Hubble telescope 125
human brain 51
human condition 82, 177, 191
haunting nature of the human condition 113
human existence 174
human experience 51, 59, 66, 68, 72
human nature 176
human umwelt 52, 59–60, 64, 68
human will 191
human genome 129
Human Epigenome Project 125
Huxley, Julian 174
hypnotism 106

I Have Been Here Before 103–105, 109
identity 178
community identity 179
doubled identity 71
erasure of identity 72
identity and victimhood 276
identity of Lancelot 59
individual identity 71
nootemporal identity 66
possibility of identity 70
ideology 207, 213n., 215, 232, 247
immortality 111
imprinting 131–134
inception 117
indexicality 38
industrialization 246

instant 51, 70
institution
  concept of institution 242
  institutionalization 242, 246, 247
integrated study of time 1, 15
intentionality
  collective intentionality 52
  short- and long-term intentionality 51–52
internal world 59, 67, 69
interruption 172, 185
irreversibility 154, 161, 162, 178
ISST 49

Jabès, Edmond 33
Jakobson, Roman 89
James, William 53, 69
Johnson, Ronald 33
Joyce, James 69, 71
Judeo-Christian tradition 192
Jung, C.G. 105, 202
justice
  destructive power of justice 276
  interplay of forgiveness and justice 274
  justice and closure 270
  justice and reconciliation 276
  justice as determination 261
  justice conducive to foregiveness 270
  justice / repayment of debts 269
  pure justice 273

Kant, Immanuel 39–41, 43–44, 141, 183
Kierkegaard, Søren 182
Kracauer, Siegfried 98
Kraus, Karl 209–210
  Benjamin and Kraus 215
King, Martin Luther 201

labor 176–178
labor camp 17
language 45–47, 49, 53, 58–60, 66, 68, 72, 181–182
  language of goals and ends 184
  metafunctions of language 54
  modelling of language 49, 53, 56
  natural language 58, 60
  origin of language 59, 72
langue (Saussure) 52
Large Hadron Collider 125
lawfulness 51
Leibnizian 43
Lévi-Strauss, Claude 44, 46, 243
life
  contemplative life 176
  lifetime 45
  non-reactionary life 199
  organic life 60
  political life 177
  vita activa / contemplativa 176
life cycle 107–8
linguistic inertia 181
linguistics
  functional linguistics 53, 54
  systemic functional linguistics 49, 53, 56
love feast 195
Löwith, Karl 172
Lucifer 34
Lucretius 34–35

M-theory
  M-theory and duration 204
magic 106, 110
Maki Yusuke 242–255
Malory, Thomas 58
Man and Time 105–106, 108, 111, 113n.–114, 121
Manhattan Project 123
Marx, Karl 172, 224n.
material processes 58
mathematics 173
matrix 35
McEwan, Ian 71
McTaggart, John 105
meaning
  experiential meaning 55
  logical meaning 55
  meaningful sequence 61–62
  organisation of meaning 59
mechanics 35, 157n., 159n.
  statistical mechanics 139–140, 154n.–155n., 158, 160
MEDEA 133, 135
medicine
  molecular medicine 128
  biomedical sciences 128
memory 47, 67, 169–170, 178–180, 184, 262
  dangerous memories 179–181
  memory and openness of time 266
  memory not repetition 268
  memory's dialogical character 266
  narrative memory 266–277
  recovered memories 266
  transforming memory 267
  traumatic memory 265

mental processes 57–58, 63, 65
mental world 52, 59, 64–66
mercury clock 105
Merleau-Ponty, Maurice 242, 246
metabolism
  carbohydrate metabolism 133
metonymy 41, 43, 46
Metz, Johann Baptist 179
Middlemarch 66–67
Mill, John Stuart 149
mimesis 85
mind 52, 57, 63
  human mind 53, 170
  minding 50–51, 60–61, 72
mitochondria 134
mode (narrative) 49, 54, 68, 70, 73
  mode of coherence 61–62, 65
modernism 68–70
  modernist narrative 69
modernity 207, 213, 218, 226
  irrationality imposed by modernity 218
  literary modernity 218, 225, 232
  literary modernity and Poe 225
  Poe's modernity 213, 226
moral worth 183
mortality 34, 213
motivation 92–93
multi-cellular organisms 125
music 214
  African American music 199
  intimidation through music 226
mystery 25, 27–33, 93–94
  classical mystery novel as a closed system 97
  double logic of mystery 97
  mystery (etymology) 26
  mystery novel seen by intellectuals 98
  mystery story 26
  mystery writing towards a misleading origin 98
mysticism 106
myth 29, 44n., 52, 61
  mythic narratives 94
mythology
  ancient mythology 105

narrative 49, 52–54, 59–60, 62, 65–73, 79–85, 87, 93, 97, 180
  constellation of story and discourse 95
  Greek narrative 63
  large patterns of narrative technique 215
  master narrative 180
  narrative advance 62
  narrative and closure 232
  narrative and forgiveness 267
  narrative focus 64
narrative mode
    of poetry 27
    of science 26
  narrative of change 172
  narrative production 64
  narrative sequence 61–62
  narrative structure of time 266
    dual temporal order of narrative 80, 81, 88
    progression from origin to endpoint 81
    progression of written narratives 80
    sequential presentation / occurrence of both tale and telling 85
    temporal loop of narrative fiction 84
  narrative studies 49, 52, 54, 62
  narrative suspense 82
  narrative technique (Poe) 225, 232
  narrative theory 52, 59–60
  narrative time 60, 229, 231
  narrative trajectory 61
  narrative transfiguring terror 229
  narrative world 62, 68, 73
  urgent narratives 265
narrator 55, 63, 67, 69, 80n.-81, 83, 85, 88, 92n., 95, 197, 201, 203, 225–226, 230–231, 265
nasic conditions 149–152
natality 169–170, 172, 175, 177–182, 187
nature
  integrative levels of nature 50
Neo-Darwinism 132
nested hierarchy 2, 8, 23
Newton, Sir Isaac 43, 47
Nietzsche, Friedrich 269
Nixon shock 240
novel
  anti-novel 73
  English novel 64
  epistolary novel 65
  postmodern novel 49
  realist novel 66
nuclear energy
  nuclear energy accidents
    Chernobyl 236, 240
    Fukushima 235–236, 239, 254– 256
    Three Mile Island 240

nuclear energy policy 236–237, 239, 242, 244, 248, 255–256
nuclear utopia 249

object
constitution of objects 39
temporal object 41
world of objects 39
Oedipus 84, 86–87, 93
oikos 177
Old Europe 233
oppression 172, 185, 190, 193, 220
oral tradition 63
order / disorder
entropy and disorder 156, 158
records of order / disorder 164
temporal order 80, 141, 185
organic world 60, 62, 64–65
origin 79–82, 84, 93, 96, 170, 175–176, 180, 185, 187
concepts of origin and future 79
confounding of origin and destination 87
dénouement as origin (Poe) 88
evolutionary origin 125
of visual pigment 130
of life 123
false origins 98
Hiroshima as origin 244, 249–250
interest in origins 123
mystery of origins 26, 29, 31–33
nature of origins 32
origin myths 262
original sin 261
origin and future 80
futures retroactively shape origins 87
origin defining future 220
virtual arc between origin and future 82
origin as a self-generative source of time 32
o. as pure becoming 32
origin in other texts 84
origin of creation 175
o. of fictional texts, events 79, 86
o. of life 133
o. of mitochondria 134
o. of order and disorder 35
o. of reading 82
o. of the universe 29–31
o. of the polis 178
o. of time 29, 31–32
origins and endpoints of texts 97
origins and evolution 123
outcome as origin 92–93
retrospection to origins 133
return to origin 267
stories of origin 260, 269
structure implied by origin and future 80
ur-origin 31
Ouspensky, P.D. 103, 106–108, 111, 114
O. and eternal recurrence 106–107
O. and time's dimensions 108
O.'s idea of the sixth dimension 117
Ouroboros 32, 203–204

paleontology 27–28
palindrome 33
Pamela 64–66
paradox 68, 139, 181
cause and effect in narrative 87
origin of time as conceptual paradox 32, 37, 40
paradox of collective identity 47
paradox of temporality 38, 40, 45, 47
paradox of Japan's nuclear energy policy 237
paradoxes of time travel 142–146
twin paradox 142–144, 148
paralogism 98
parapsychology 110
parole (Saussure) 52
passive resistance 200
past/s 178–181, 184, 186
ability of the past to form the future 267
closing the past 276
dismissal of the past 262
fore-given past 261
healing the past 272
investigations of the past 27
legacy of the past 178
new past emerging in forgiveness 267
patterns of the past 272, 277
struggle with the past 108
transformative illumination of the past 268
trauma recreating the past 266
pendulum clock 105
performative 38
permanence 178
person 41–42
restoration of the person 275
phase space 156, 158–159, 163
phenomenology 39–40, 42–44

philosophy 173, 176, 180, 186
  philosophy of history 172
  Western philosophy 176
photosphere 30
physical change 50
physical world 58, 59–63, 65–67, 69, 72
physics 49, 52, 106
Planck telescope 125
Plato 173, 176–177
Platonic trap 176
plot 26, 84–85, 87–88, 90–94, 96–97, 207–208, 215, 224, 266
  plot dynamic 83
Poe, Edgar Allan 88–90, 96–97
  Europe's appreciation of Poe 233
  Poe and American critics 232
  Poe and crowds 226
  Poe and literary modernity 214
  Poe and terror 229–230
  Poe's literary principles 214, 218, 224
  Poe's modernity 213
poetic principle 89
polis 177–178, 182
politics 176–178, 183, 186, 209–210, 218, 221–222, 233, 271
Popper, Karl 149
post traumatic stress syndrome 15
Post-Genome era 125
postmodern, postmodernism 49, 53, 68, 70–71
power of pause 198
practical reason 183
pragmatics 54
precognition 105–106, 111–113, 115
prediction 51, 151
presence-at-hand 37
present 51, 67, 179–180, 184, 186
  endless present 265
  noetic present 174
  social present 171
Priestley, J.B. 103–106, 108
  Priestley as a dramatist 110
  Priestley compared to Fraser 121
  Priestley on J.W. Dunne 110
  Priestley on precognition 112
  Priestley the time-haunted playwright 120
printing 63–64
process
  process types 57–59
  relational processes 58
progress 66, 80, 124, 158, 180, 183–184, 186, 207, 209, 215–216, 218–219, 222, 229–230
progression 64, 80–81, 85, 90, 93, 116
  semblance of progression 86
prolepse 87
promise 169, 180–182, 185–186
  promise-maker 181
propaganda 16, 209
Proteome 125
Proust, Marcel 52
pura vida 199
Pynchon, Thomas 71

quantum 154–155, 157
  quantum physics 105
  quantum theory 52
  quantum vacuum 32
  quasi quantum vacuum 33

race problem 191
radiation 235–236, 254, 256
radioactive material 51, 236, 256
realism 92, 96
  realism as camouflage 92
  realist camouflage of structure 94
  realist novel 66
  realistic literature 91
  realistic surface vs. thematic structure 97
reality 173, 184
  fictional reality 68
reciprocity
  male-female reciprocity 196
reconciliation 259–260, 263, 264, 266, 269–277
  reconciliation creating a mutually sustainable future 271
  group reconciliation 272
  reconciliation must be dialogical 271
  reconciliation opposed by traumatic persistence 275
  visibility of reconciliation 271
redemption
  redemptive ending of fairy tales 83
reincarnation 107
relativity (physics), see Theory of Relativity
relativity of being 40
religion 176, 195–196
repentance 260, 263–264
repetition 32, 47, 88, 107, 223, 226, 228, 263, 267–268
  a-temporality of repetition 265
  absolute repetition 107
  damning closure of repetition 277
  forgiving is release from repetition 271

repetition and trauma 265
repetition of offenses 262
representation 39–40, 45, 79–80, 85, 91, 141, 143–144, 189, 192, 203, 212–215, 219–220
representation of consciousness 63–65
symbolic representation of reality 37
resentment 268, 270
retrodiction 151
retrospection 123, 125, 128, 135
retrospection to origins 133
retrospective emphasis 124
retrospective obsession 123
revelation 267–268, 277
time of revelation 267
revenge
revenge fantasy 199
reversals of origin and outcome 84
reverse construction (Poe) 89
revolution
chronometric revolution 27
permanent revolution 174
rhythm 89n., 190–193, 195–196, 199, 201, 204
rhythm disruptions 198–199
Richardson, Samuel 64–65
Ricoeur, Paul 53
ritual 52, 61
Robinson Crusoe 64
romance 59
Roosevelt, Franklin D. 18
rupture 172, 182, 185, 207, 222
eschatological rupture 172

salvation 260–261
Sartre, Jean-Paul 88
Saussure, Ferdinand de 52
Sayers, Dorothy 93, 98
Sayers on Aristotle 93
Scheler, Max 40
science 5, 11, 26, 30–31, 34, 49, 70, 172–173, 180
applied science 123–124
basic science 124, 126, 130, 135
programmatic advancement of science 124
scientific progress 124
science projects' temporal orientation 126
science fiction 134
Science (journal) 127–128
Sebeok, Thomas 98
secularization 172
self-purification 200
semantic/s 54, 55
semantic acts 58
semantic fuzziness 57
semantic metafunction 55
semantic process 56
semiotic/s 53, 55, 58–59
semiotic media 53
sequence 61–63, 65–69, 80–82, 85–87, 91–92, 97
associative sequence 64
chronological sequence 64–65
equative sequence 62
logic of sequence 84
sequence of reading 81
temporal sequence 32, 91–92
serialism 110
serialism with infinite regress 110
Serres, Michel 35
sex 106
Shakespeare, William 127, 207, 214
Sherlock Holmes 84, 94–95
simultaneity in time 193
sjužet 53, 80
Slaughterhouse-Five 189
slavery 195, 197
slavery an outside symbol 197
slavery conventions 195
slavery patterns 194
US slavery experience 194
social codes 59
social justice 105
social oppression 190
social practice 44, 45
social world 52, 59–60, 62–63, 65–68
socialism 110
society 52–53
consumer society 178
Sophocles 84, 86
soul 107, 111, 169, 176
eternal soul 176
liberation of the soul 173
soul-enhancing possibilities 107
space 108, 116, 125
space exploration 125
space-time 53, 69, 114, 116
spirit
eternal spirit 169
spreading 157
future spreading 164
past spreading unchangeable 164
spatial and temporal spreading 158
spreading as a primitive clock 156

spreading as descriptor of entropy 156
spreading is sequential 163
spreading of energy states 158
spreading, past and present 159, 163
Stalin 18, 22, 216, 226–227, 229
Stalingrad 227
Battle of Stalingrad 226, 232
stasis
stasis of bliss 261
stasis of same-again 261
state 177
Stein, Gertrude 84
stochastic 50–51, 152
structuralism 37, 44, 52
sub-atomic theory 117 n. 8
subject
content and the object of consciousness 46
empirical subject 39
extratemporal subject 39–40, 43
transcendental subject 39
subjectivity 39, 41–42
subversive writer 95
Sulston, John 127
sundials 105
superman 106
surrealism 217–218, 220
"Surrealism" (Benjamin essay) 216–218
suspense 81–82, 87, 90
symptomatic drama 266
synchronicity 106
System of Transitivity 56

tarot 106
telos 10, 235, 237, 244, 249–250, 252, 254–256
teleology 47, 186
teleological continuity 172
teleological revision 135
temporal form 42–43
temporality 4, 5, 23, 26, 37, 40, 42, 50–51, 53, 59, 61, 68–69, 71, 73, 170–171, 190
atemporality 68, 70, 170
biotemporality 51–52, 60, 62, 65, 68, 170
cyclical temporality 27
eotemporality 51, 68–69, 170
experience of temporality 43
human temporal condition 169, 172, 182
inversion of temporal structure 266
linear, causal model of time 25–26
linear time 32
narrative temporality 60
nootemporality 52–53, 65, 67–69, 170–171, 173, 182, 184
paradox of temporality 37, 40
prototemporality 51, 68, 70, 170
sociotemporality 52, 60–62, 68, 170–172
temporal condition 169, 172, 175, 182, 184
temporal dynamics 108
temporal level 182
temporal orientation 135
temporal structure 261, 266
temporalities 49–50
temporality of truth 44
tenor 54–55
terror
aesthetics mastering terror 232
formula linking time and terror 229
historiography and fiction as means to overcome terror 215
law of terror in Poe 230
master of terror, master of time 208, 229
terror and time 228
terror clouding the past 211
terror and waiting 208, 227, 232
terrorism 208–211
damage inflicted by terrorism 211
discourse of terrorism 209
duration of terrorist campaigns 211
international terrorism 210–211
terrorism and time 227–228
terrorism defined by impossibility of victory 211
texture
narrative texture 69
theatre 103–104, 107, 109–110, 117, 120, 253
conventional London theatre 120
theatre industry 109
Theory of Relativity 52–53
General Theory of Relativity 148 n. 4
Special Theory of Relativity 140–144
consequences of Special Theory of Relativity 148
Special Theory of Relativity and atemporality 144
Special Theory of Relativity and the branching tree model 152–153
Special Theory of Relativity and travel into future 142
thermodynamics
entropy relation in thermodynamics 139–140

First Law of Thermodynamics 164
irreversibility a feature of the Second Law of Thermodynamics 154
thermodynamic probability 158–159, 162
thermodynamic probability as criterion for the anisotropy of physical time 162
Second Law of Thermodynamics 151–152, 154–156, 160–161, 164

Thomas, Dylan 24

Three Nuclear Principles (Japan) 248

time
absolute time 111–112
arrow of t. 25, 173, 184
artistic images of time 105
canonical forms of time 170
common-sense idea of time 106
deep time 193
discontinuities as centers of time 191
dimensions of time 111, 120
second time dimension 111
uni-dimensional time 111
dreamtime 116
eschatological (model of) time 172, 244
evolution of time 170, 173, 175
experience/s of time 105
female time-experience 198
flow of time 184–185
integrated study of time 2, 15
lifetime 38
linear model of time 25–26
measurements of time 105
multi-dimensionality of time 113
narrative structure of time 266–267
non-linear view of time 104
origin of time 29, 31–32, 38–39, 40, 41, 42, 44–45, 47
perceived time 105
possibility of time 51
productive experience of time 191
pure time 8–9, 189–191, 197–200, 203
pure-time generation 202
quantitative time 61
reality of time 184
social time 45
space-time 53, 69, 114, 142, 144–145, 147, 148n., 155n.
spatialized time 111, 197
spiral track of time 104
structure of time as opening 268
time as code 42–43
time the principle of translation 43
time felt 2, 60–61
theory of time
hierarchical theory of time 170
theory of time's conflicts 170, 172–173
third time dimension 112
time as an order of eternal recurrence 106
time a curved continuum 203
Time and Free Will (Bergson) 190–191, 199, 202
time as linear, future-directed impetus 192, 204
time as monovalent concept 62
time as separate from and disdainful of the material world 192
time as singular noun 49
time consciousness 111
awareness of passing 169
time division 105
time of Nature/God 243
time smith 170
time understood 15, 60–61
Time vs. Thing 192
time leap 113
time recurrence 103–104, 106
time split 115
time theories 103–104, 115
time track 109
time-outside-of-time 117, 190
timeless 173
transfiguration of time 267
trick of time 104, 114
understanding of time 49, 52
ultimate time 112
Western conception of time 191
world-time 38–39, 45

Time and the Conways 104–105, 110, 112

time imagery 237, 242–244, 255–256
circular time 244
four ideal types of time imageries 243
modern linear time 244, 252, 256
aimless linear time 255
oscillating time 243, 253, 255
segmented linear time 244, 249, 255–256
Fukushima accident shaking dominant postwar time imagery (Japan) 255

Time Plays (Priestley) 104, 110, 121

time travel 116, 120, 139, 142, 155, 160, 163
cosmological time travel 145
paradoxes of time travel 145
time travel according to Wells 142
time travel and entropy balance 146

time travel and Kant 141
time travel and scattered stages of existence 147
time travel and the twin paradox 142
time travel from the future 144
time travel into the future 142–144
time travel into the past 144–145
time travel into the past impossible 163
Todorov, Tzvetan 52
transcendence 201
Transcriptome 125
translation 43–45
transfiguration 262, 267–268
trauma 259, 265–266, 271, 273, 275–277
mass trauma 275
post-trauma unconsciousness 117
sacralized trauma 275
time of trauma 265
traumatic persistence 275–276
Trotsky, Leo 174
Truth and Reconciliation Commission 264
twin paradox 142–144, 148

umwelt 50–52, 59–60, 64, 68–70, 72, 170–172, 182, 186
nootemporal umwelt 172
sociotemporal umwelt 171
temporal umwelt 174, 182
Ursprung 187
utility 127
utopia 244
utopian visions 172

vengeance 178–179, 203
verb 57–58, 63
verbal processes 55–57, 63, 80
victim
victims of injustice 184
victimhood 267, 275–276
visual pigment 130
virtue 59
vortex 34–35
Vonnegut, Kurt 189, 203

war 15, 20, 210–211, 216, 226–227, 233, 235, 244–247, 249–250, 255, 271
water clock 105
Watson, James 127
Webb Telescope 126
Wells, H.G. 105, 142–145, 163
What Dreams May Come 117
Whitrow, G.J. 23
Wiesenthal, Simon 272–273
Wittgenstein, Ludwig 38
Wolf, Fred Alan 189
women's rights 183
Woolf, Virginia 69
work 176–178
work of transformation 180
World War I 16, 109, 113
World War II 15, 112, 118, 120, 246

yoga 106

Zeitgestalt 42

www.ingramcontent.com/pod-product-compliance
Lightning Source LLC
LaVergne TN
LVHW020651300926
849161LV00004B/902
* 9 7 8 9 0 0 4 2 5 1 6 8 7 *